A Compendium of General Systems Theory 2.0

-- The Structure-Behavior Coalescence Approach --

William S. Chao

Structure-Behavior Coalescence

$$\text{Systems Architecture} = \text{Systems Structure} + \text{Systems Behavior}$$

4

CONTENTS

ABOUT THE AUTHOR

Dr. William S. Chao is the CEO & founder of SBC Architecture International®. SBC (Structure-Behavior Coalescence) architecture is a systems architecture which demands the integration of systems structure and systems behavior of a system. SBC architecture applies to hardware architecture, software architecture, enterprise architecture, knowledge architecture and thinking architecture. The core theme of SBC architecture is: Architecture = Structure + Behavior.

William S. Chao received his bachelor degree (1976) in telecommunication engineering and master degree (1981) in information engineering, both from the National Chiao-Tung University, Taiwan. From 1976 till 1983, he worked as an engineer at Chung-Hwa Telecommunication Company, Taiwan.

William S. Chao received his master degree (1985) in information science and Ph.D. degree (1988) in information science, both from the University of Alabama at Birmingham, USA. From 1988 till 1991, he worked as a computer scientist at GE Research and Development Center, Schenectady, New York, USA.

Dr. William S. Chao has been teaching at National Sun Yat-

Sen University, Taiwan since 1992 and now serves as the president of Association of Enterprise Architects, Taiwan Chapter. His research covers: systems architecture, hardware architecture, software architecture, enterprise architecture, knowledge architecture and thinking architecture.

PART I: GENERAL SYSTEMS
THEORY 1.0

12

The Need of Systems Definition

The need for defining a system arises because any real-life system is inherently complicated. It is impossible to comprehend fully the intricate interaction of any system of the real world with its environment, or to describe all its components and each of its details.

Systems definition is an "artifact" created by humans to describe what a system is.

Without a systems definition, everybody has his own saying about a system and never be able to reach a consensus.

General Systems Theory 1.0 Defining a System

General systems theory 1.0 defines a system, hopefully to be an integrated whole, embodied in its components, their interrelationships with each other and the environment, and the principles and guidelines governing its design and evolution.

As a first example, general systems theory 1.0 defines a *Fish*, hopefully to be an integrated whole embodied in its assembled components of *Body* and *Fin*, their interrelationships with each other and the environment, and the principles and guidelines governing its design and evolution.

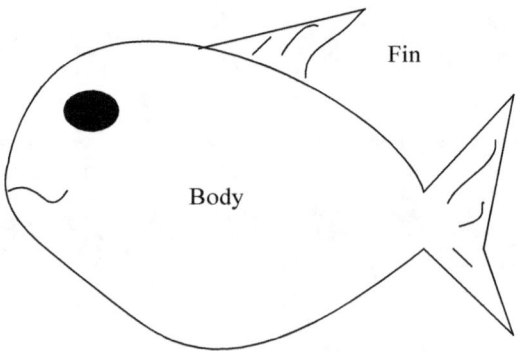

As a second example, general systems theory 1.0 defines *Earth*, hopefully to be an integrated whole embodied in its assembled components of *Land* and *Oceans*, their interrelationships with each other and the environment, and the principles and guidelines governing its design and evolution.

As a third example, general systems theory 1.0 defines the *Wardrobe_A*, hopefully to be an integrated whole embodied in its assembled components of *Drawer_1*, *Drawer_2* and *Drawer_3*, their interrelationships with each other and the environment, and the principles and guidelines governing its design and evolution.

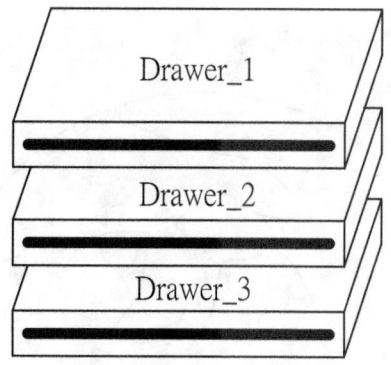

As a fourth example, general systems theory 1.0 defines an *Eyeglasses*, hopefully to be an integrated whole embodied in its assembled components of *Frames* and *Lenses*, their interrelationships with each other and the environment, and the principles and guidelines governing its design and evolution.

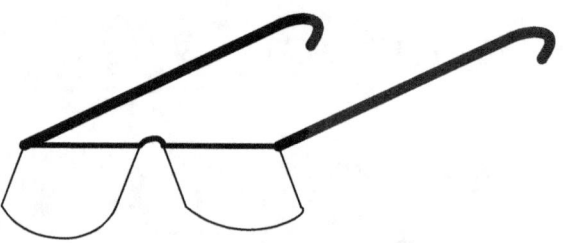

As a fifth example, general systems theory 1.0 defines a *Swing*, to be hopefully an integrated whole embodied in its assembled components of *Ropes* and *Seat*, their interrelationships with each other and the environment, and the principles and guidelines governing its design and evolution.

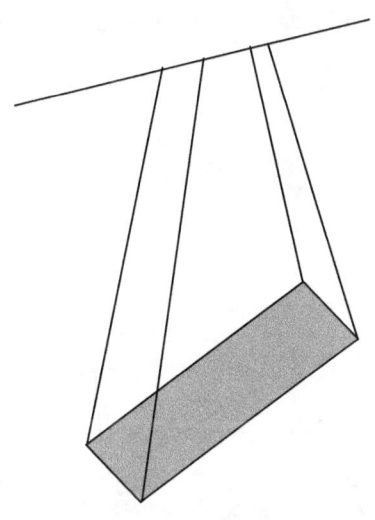

Physical and Virtual Systems

A physical system exists in the physical world. A physical system is also called a concrete or real system. For example, a *Bicycle* composed of *Wheels*, *Frame* and *Pedal* is a physical, concrete, or real system.

A virtual system is a system that is composed of non-physical components, i.e., ideas, thoughts, or notions. A virtual system exists in the virtual, abstract, or notional world. For example, a fairy tale *"Jack and the Beanstalk"* composed of *"Jack"* and *"the Giant"* is a virtual, abstract, or notional system.

Boundary and Environment of a System

We scope a system by defining its boundary. All components of the system are inside the boundary while the environment is outside the boundary.

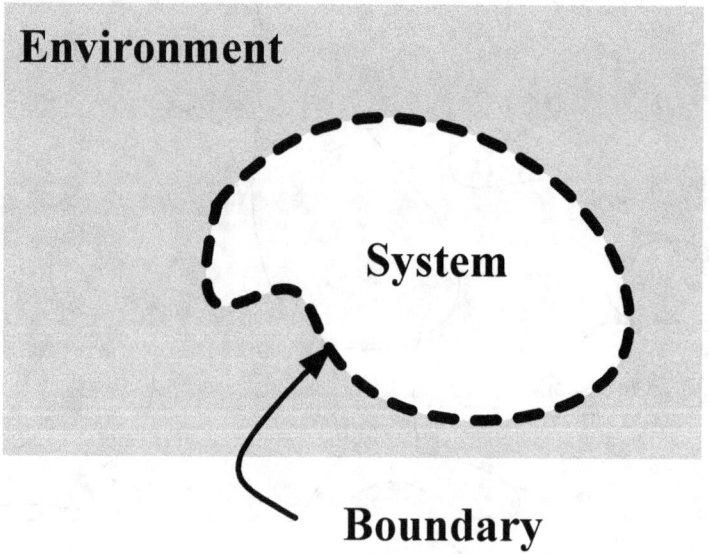

The environment is also known as the surroundings. A system may or may not interrelate with the environment. An open system interrelates with the environment through the exchange of matter, energy, data, information, or message.

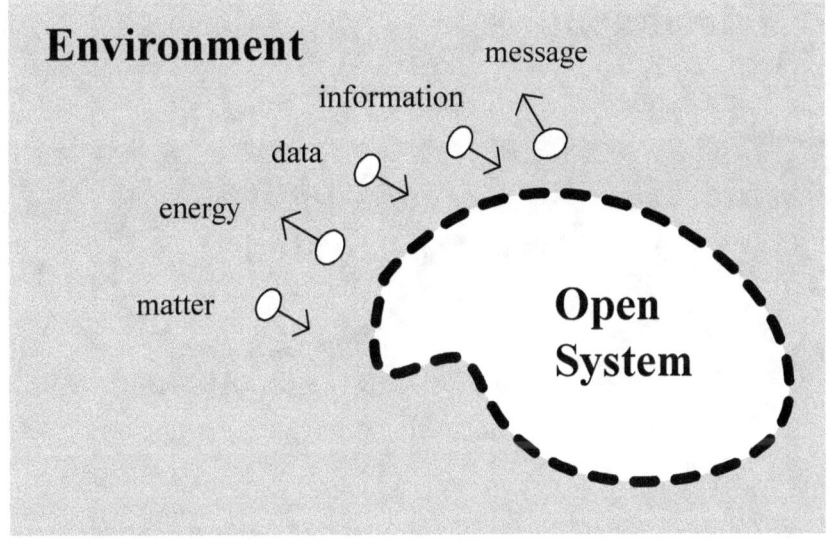

An isolated system does not interrelate with the environment at all. There is no exchange of matter, energy, data, information, or message between the isolated system and the environment.

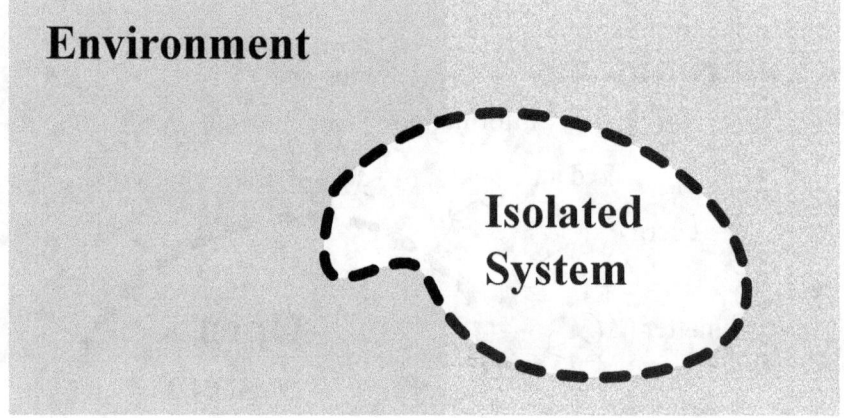

Higher-Order Systems

Higher-order systems interrelate with the environment through the exchange of not only matter, energy, data, information, or message but also systems.

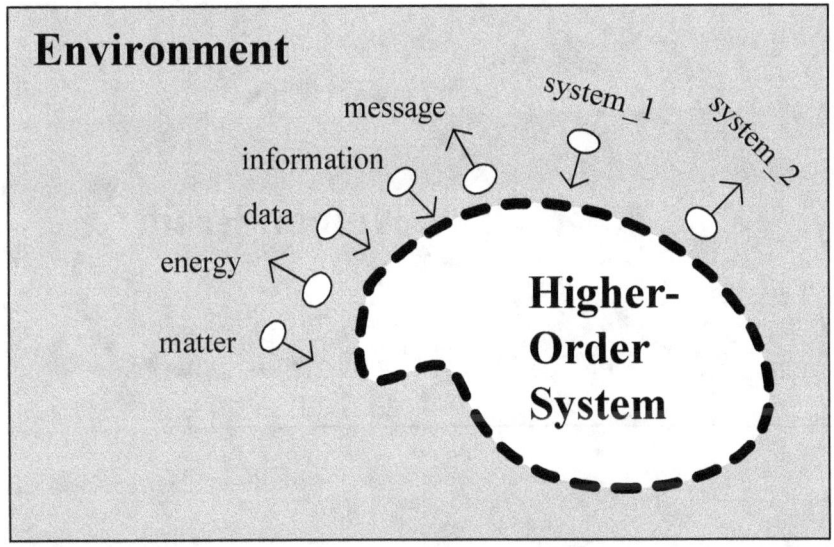

Human brain is regarded as higher-order systems. The human brain is a higher-order system, because it is able to produce a large number of systems in which *System_1*, *System_2* and *system n* are the output of the human brain.

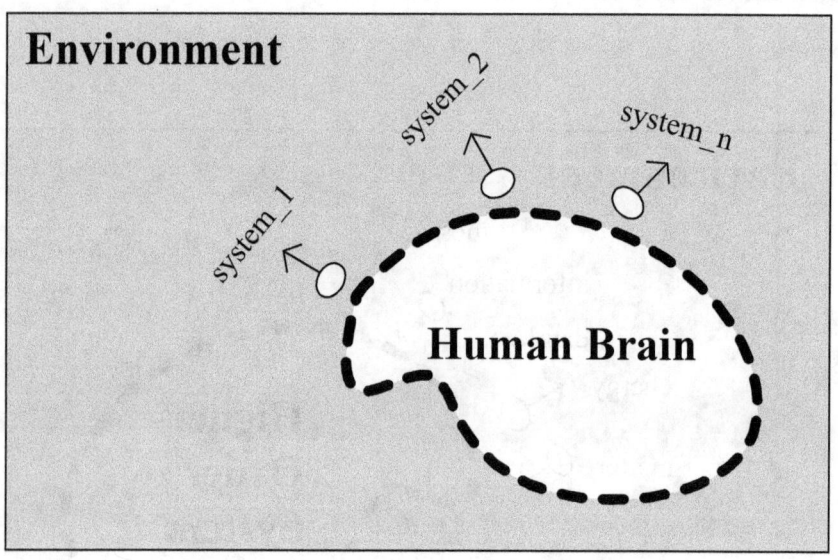

Strategic management is a higher-order system. Strategic management, for each strategy will output a system (goal), as shown below. In the figure, *Strategy_1* and *Strategy_n* are the input of the strategic management; *System_1* and *System_n* are the output of the strategic management.

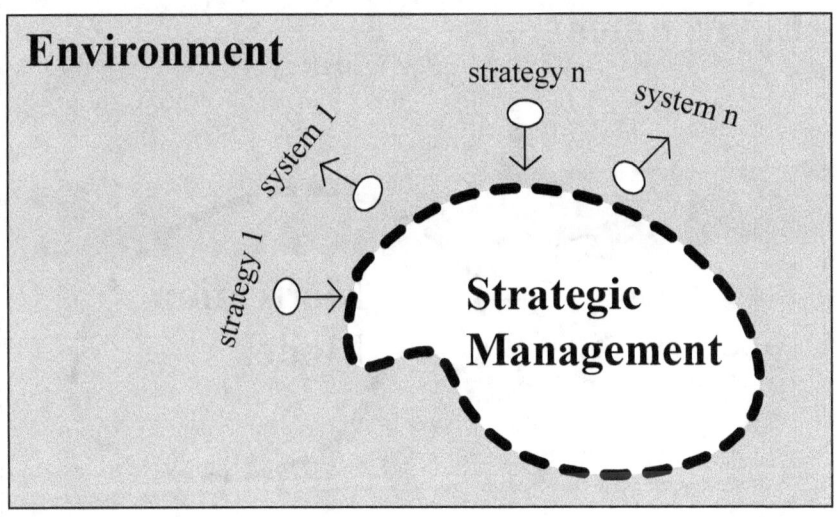

Motivation model is also regarded as a higher-order system. Motivation model will output a system (goal) for each strategy as shown below. In the figure, *Strategy_1* and *Strategy_n* are the input of the strategic management; *System_1* and *System_n* are the output of the motivation model.

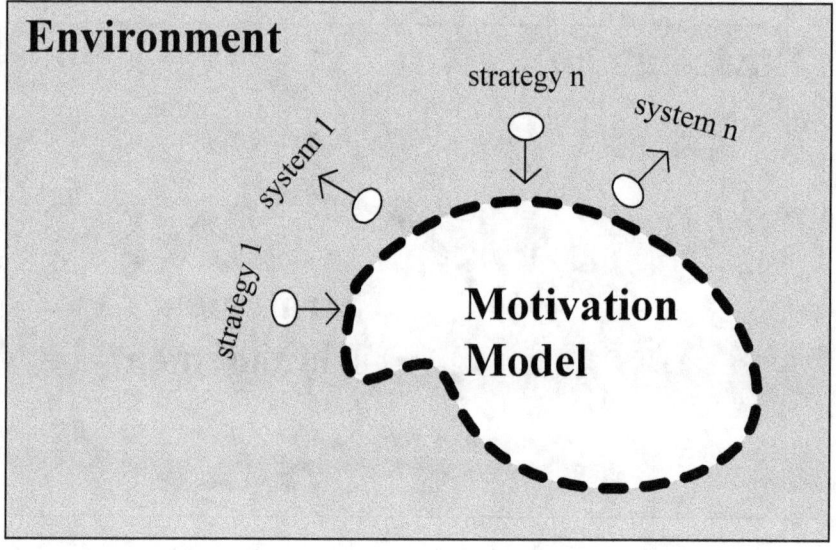

System dynamics is also regarded as a higher-order system, because it dynamically simulates the causal relationship among a large number of systems such as *System_1*, *System_2* and *System_n* as shown below. From these simulated systems, a decision maker thus is able to strategically choose the most appropriate one.

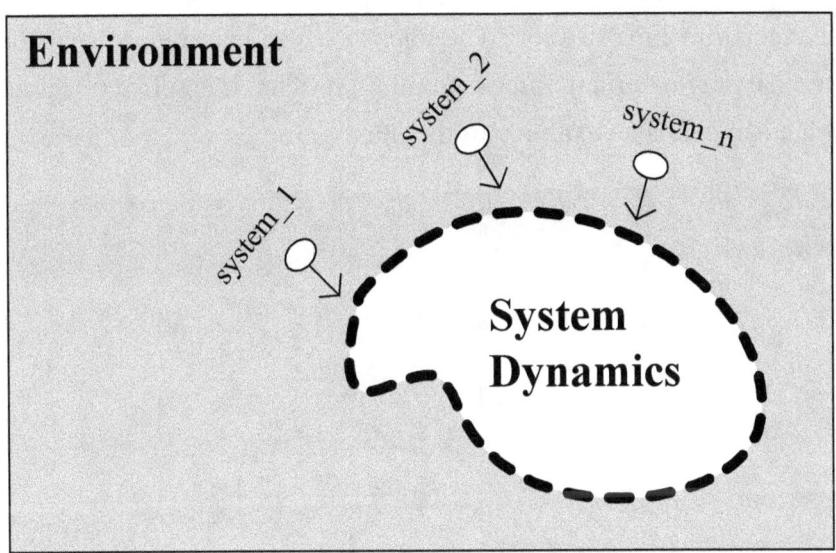

Evolution of a System

A system, not matter it is physical or virtual, will always change from time to time. The change cause may come from the internal or external forces of the system. A self-replicating organism cell is an example of the internal forces.

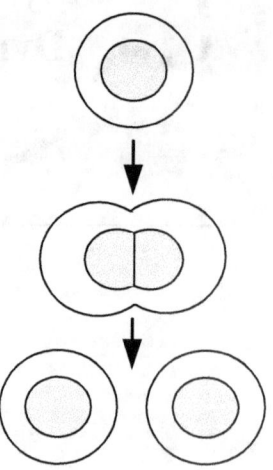

A worker reshaping, rebuilding, or remodeling a system is an example of the external forces.

A system evolves when it changes. Evolution of a system is shown as follows.

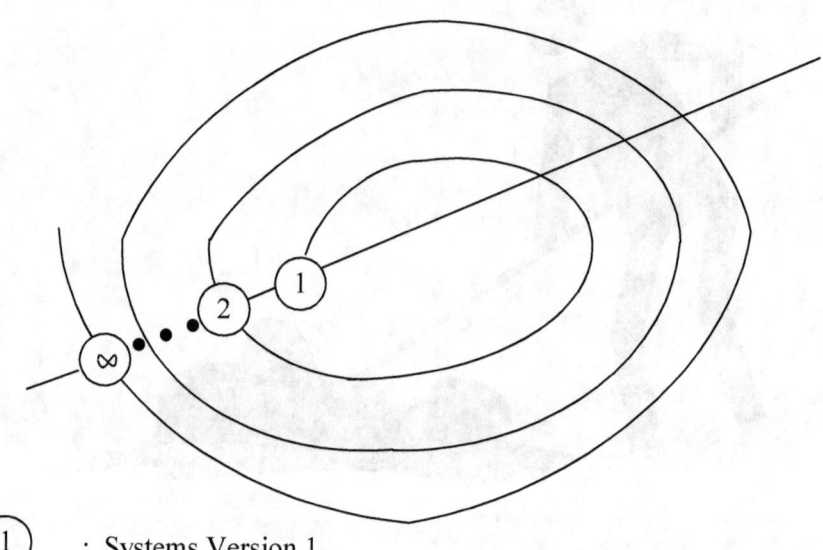

① : Systems Version 1

② : Systems Version 2   • • •   ∞ : Systems Version ∞

As an example, the following figure shows the general systems theory 1.0 systems definition *version 1* defining the *House_B* to be hopefully an integrated whole embodied in its assembled components of *Roof_1*, *Window_1* and *Door_1*, their interrelationships with each other and the environment, and the principles and guidelines governing its design and evolution.

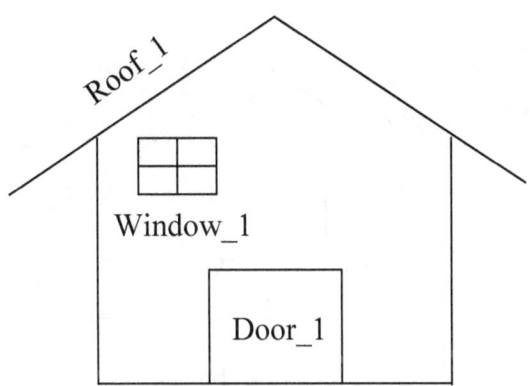

After the *House_B* changes and evolves, the following figure shows the general systems theory 1.0 systems definition *version 2* defining the *House_B* to be hopefully an integrated whole embodied in its assembled components of *Roof_1*, *Window_1*, *Window_2* and *Door_1*, their interrelationships with each other and the environment, and the principles and guidelines governing its design and evolution.

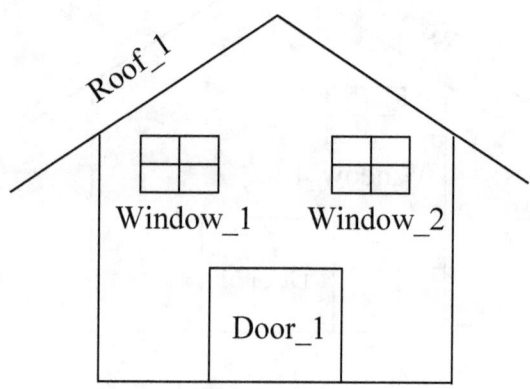

PART II: SHORTCOMINGS OF GENERAL SYSTEMS THEORY 1.0

General Systems Theory 1.0 Does Not Define the Integration of Systems Structure and Systems Behavior

Systems structure and systems behavior are the two most significant views of a system. In order to achieve a truly integrated whole of a system, we first need to integrate the systems structure and systems behavior together.

In other words, integration of systems structure and systems behavior results in the integration of a whole system.

Since general systems theory 1.0 does not define the integration of systems structure and systems behavior, very likely it only hopes and will never be able to truly form an integrated whole of a system.

General Systems Theory 1.0 is Powerless in Defining a System Appropriately

General systems theory 1.0 does not define the integration of systems structure and systems behavior, very likely it only hopes and will never be able to faithfully form an integrated whole of a system.

In this situation, general systems theory 1.0 is powerless in defining a system appropriately.

PART III: INTRODUCTION TO SYSTEMS ARCHITECTURE

Multiple Views of a System

In general, a system is extremely complex that it consists of several evolution&motivation views such as strategy/version n and strategy/version n+1 views; it also consists of various multi-level views such as concept, analysis, design and implementation views; it also consists of many systemic views such as structure, behavior and input/output data views.

In a system all these strategy/version n, strategy/version n+1, concept, analysis, design, implementation, structure, behavior and input/output data views represent the multiple views of a system.

40

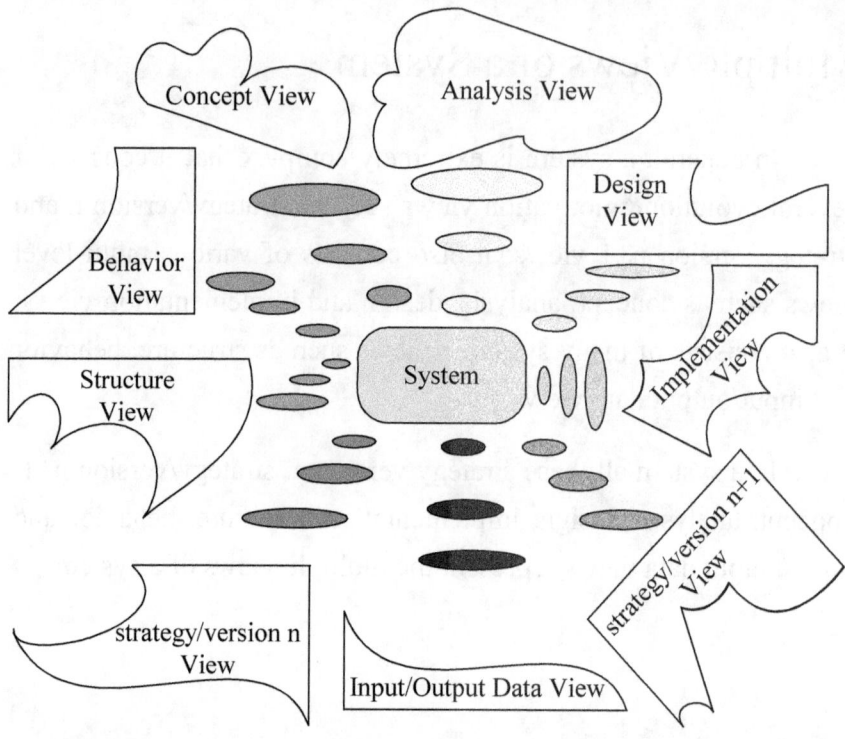

Systems Model

A systems model (SM) is a virtual system, distinguished from a physical system, used to describe and represent either the physical or virtual systems.

Below shows a physical system in which there are two buildings located in the upper left side and right underneath. The upper left building is Jackson Hotel and the right underneath building is Clinton Theater.

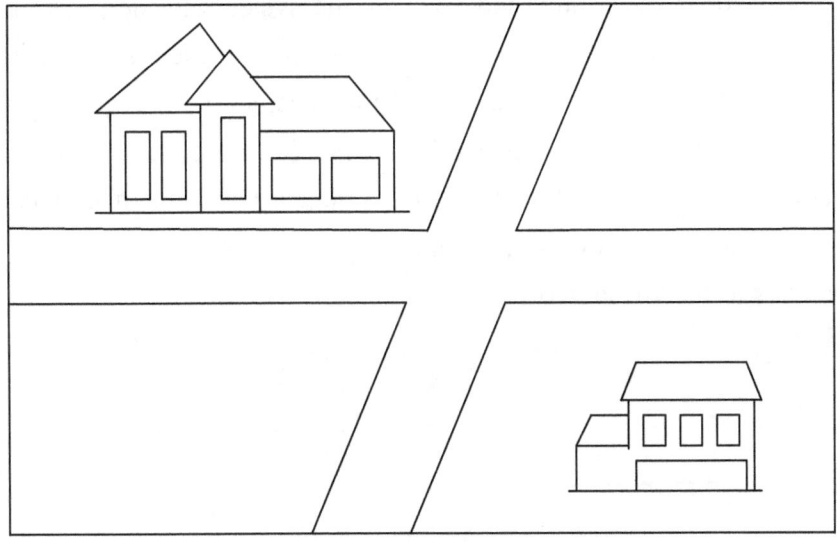

To model the physical system shown above we may then obtain a map as shown below. The map is a kind of systems model used to describe and represent the physical system.

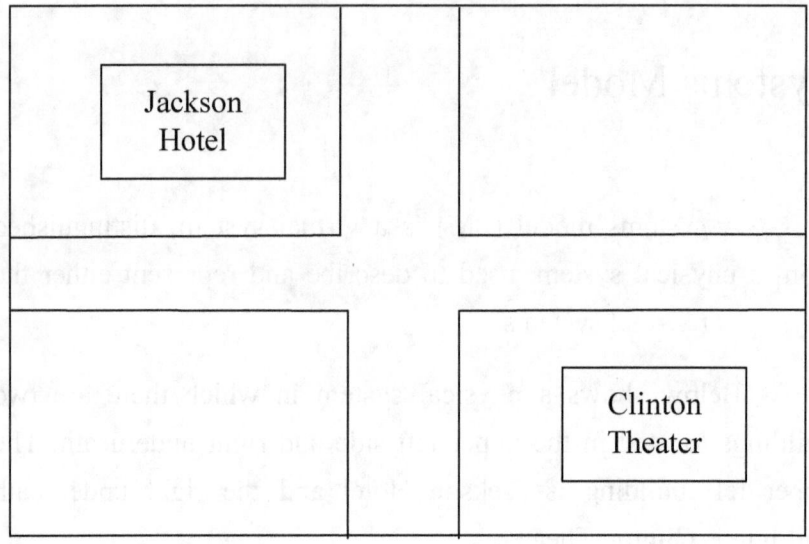

     Besides describing and representing systems in the physical world, a systems model can also describe and represent systems in the virtual world. The virtual world includes a software system, a virtual reality, or a thought within a person's mind, etc. As shown below that a fashion designer is designing a new suit of clothes. Designing a suit of clothes, being a thought inside a person's mind, belongs to the virtual world.

To model the thought within a person's mind shown above, we may then use a clothes design diagram as shown below. The clothes design diagram is a kind of systems model used to describe and represent a person's thought.

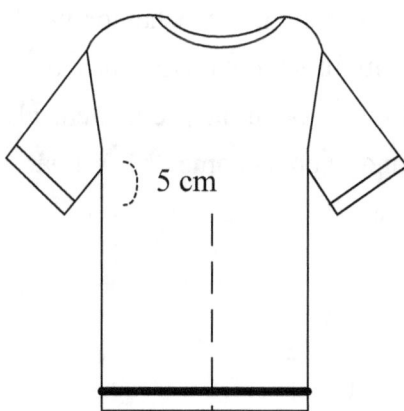

Non-Architectural Approaches Versus Architectural Approaches

The systems model describes and represents the system multiple views possibly using two different approaches. The first one is the non-architectural approach and the second one is the architectural approach.

The non-architectural approach, also known as the model multiplicity approach, respectively picks a model for each view as shown below, the strategy/version n view has the strategy/version n model, the strategy/version n+1 view has the strategy/version n+1 model, the concept view has the concept model, the analysis view has the analysis model, the design view has the design model, the implementation view has the implementation model, the structure view has the structure model, the behavior view has the behavior model and the input/output data view has the input/output data model. These multiple models are separated, always inconsistent with each other, and then become the primary cause of model multiplicity problems.

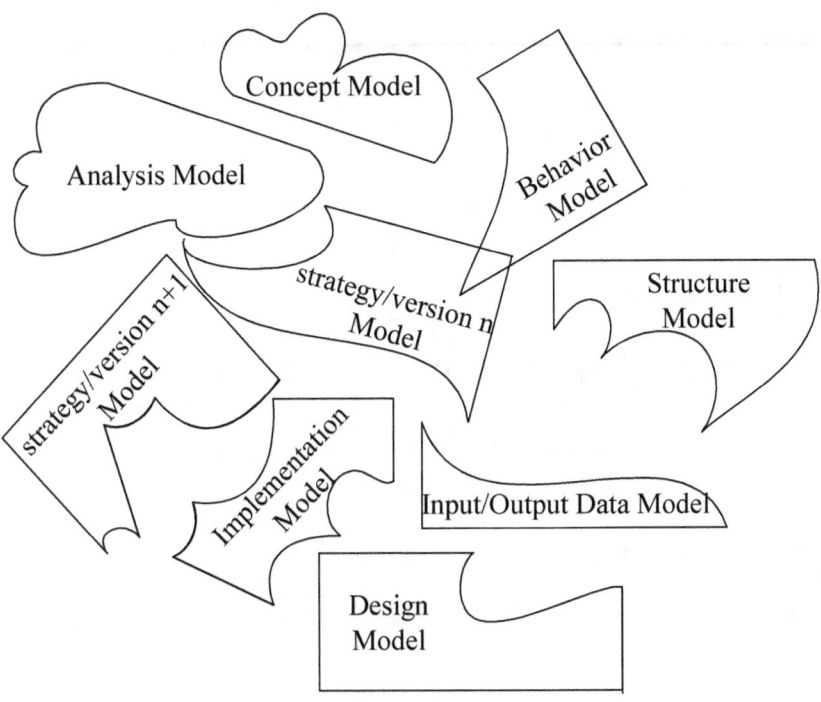

The architectural approach, also known as the model singularity approach, instead of picking many heterogeneous and unrelated models, will use only one single coalescence model as shown below. The strategy/version n, strategy/version n+1, concept, analysis, design, implementation, structure, behavior and input/output data views are all integrated in this multiple views coalescence (MVC) model of systems architecture.

Definition of Systems Architecture

Involved systems are extremely complex in every aspect so that each stakeholder needs a blueprint or model to capture their essential structures and behaviors. Systems architecture is such a blueprint or model.

We give systems architecture a definition of our own as shown below.

Systems architecture is an integrated whole of a system's multiple views, i.e., structure, behavior, and other views, embodied in its components, their interactions with each other and the environment, and the principles and guidelines governing its design and evolution.

From the above definition, we find out that systems architecture is an integrated whole of a system's multiple views, i.e., structure, behavior and other views, embodied in its assembled components, their interactions (or handshakes) with each other and the environment, and the principles and guidelines governing its design and evolution. That is, systems architecture is an integrated and coalescence model of multiple views. In this coalescence model, structure, behavior and other views are all included in it as shown below. We do not supply each view a respective model in this systems architecture coalescence model.

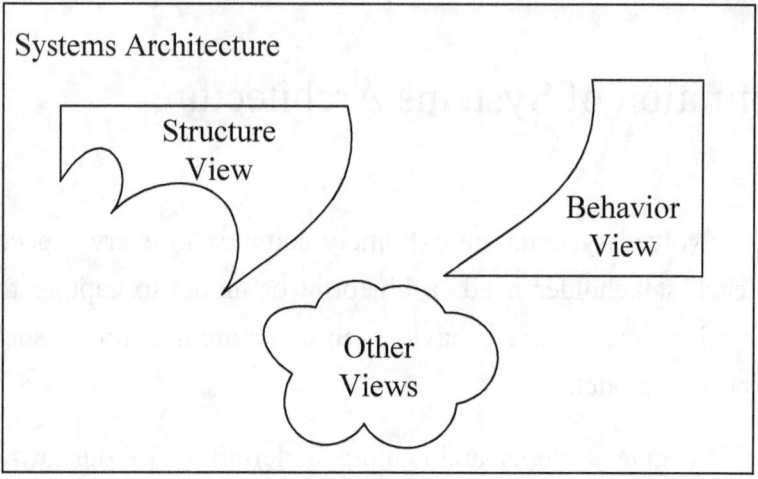

Since multiple views are embodied in a system's assembled components which belong to the structure view, they shall not exist alone. Multiple views must be loaded on the structure view just like a cargo is loaded on a ship as shown below. There will be no multiple views if there is no structure view. Stand-alone multiple views are not meaningful.

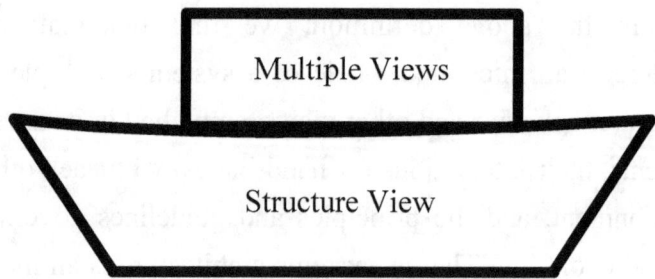

Architecture Description Language

An architecture description is a formal description and representation of a system. A description of the systems architecture has to grasp the essence of the system and its details at the same time. In other words, an architecture description not only provides an overall picture that summarizes the whole system, but also contains enough detail that the system can be constructed and validated.

The language for architecture description is called the architecture description language (ADL). An ADL is a special kind of language used in describing the architecture of a system.

Since the architectural approach uses a coalescence model for all multiple views of a system, the foremost duty of ADL is to make the strategy/version n, strategy/version n+1, concept, analysis, design, implementation, structure, behavior and input/output data views all integrated and coalesced within this architecture description.

Systems Architecture as a Knowledge Repository

Based on its definition, systems architecture can be regarded as a knowledge repository of a system. Each stakeholder, through structure, behavior and other views, submits his own knowledge and expertise to this repository when the systems architecture is built up, as shown below.

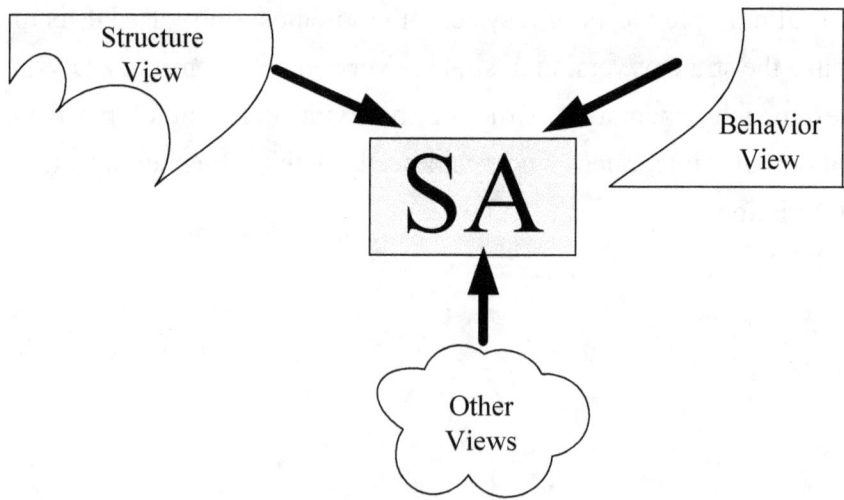

On the other hand, any stakeholder, if there is any request then he would query the system architecture. The result of the query is gathered into a view for stakeholders to see or read, as shown below.

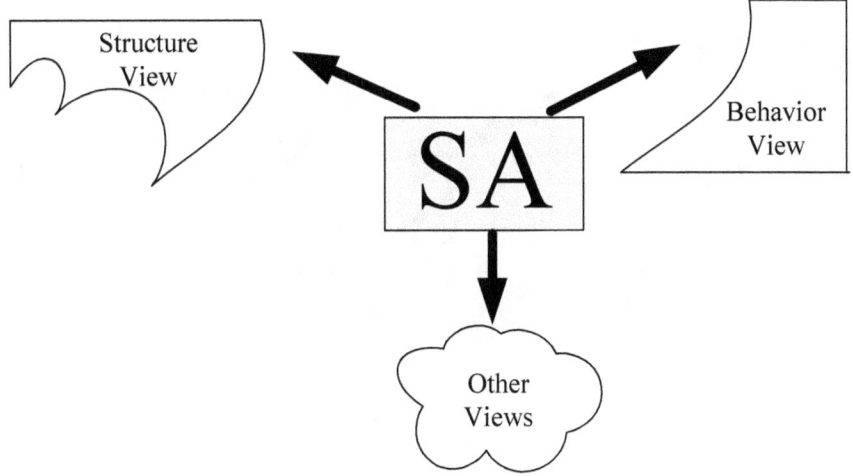

Combining the above two figures, below figure tells us that systems architecture is exactly a knowledge repository of a system. Stakeholders can submit and acquire knowledge to and from the systems architecture.

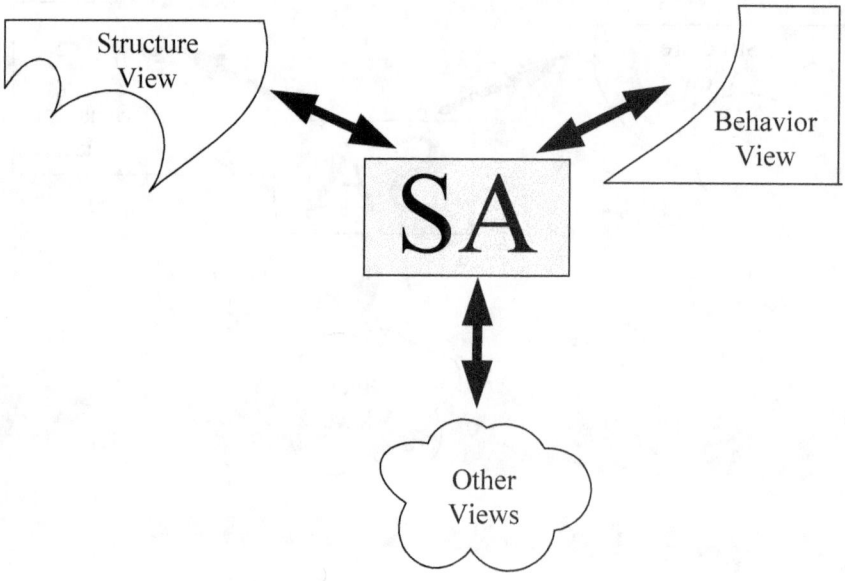

When used as a knowledge repository of a system, systems architecture becomes a communicating tool for comprehension enhancement, internal collaboration and interworking with partners. The systems architecture also supplies documented systems structures and systems behaviors.

Constructing the Systems Architecture Iteratively and Evolutionally

Systems architecture shall not be constructed in one step. On the contrary, a systems architect must construct the systems architecture iteratively and evolutionally. Iterations and evolutions allow systems architects to demonstrate incremental values of their works and obtain early feedback of the systems architecture.

Below figure shows that the systems architecture *version 1*, *version 2*, *version 3*, *version 4*,..., and *version* ∞ are constructed iteratively and evolutionally by a systems architect.

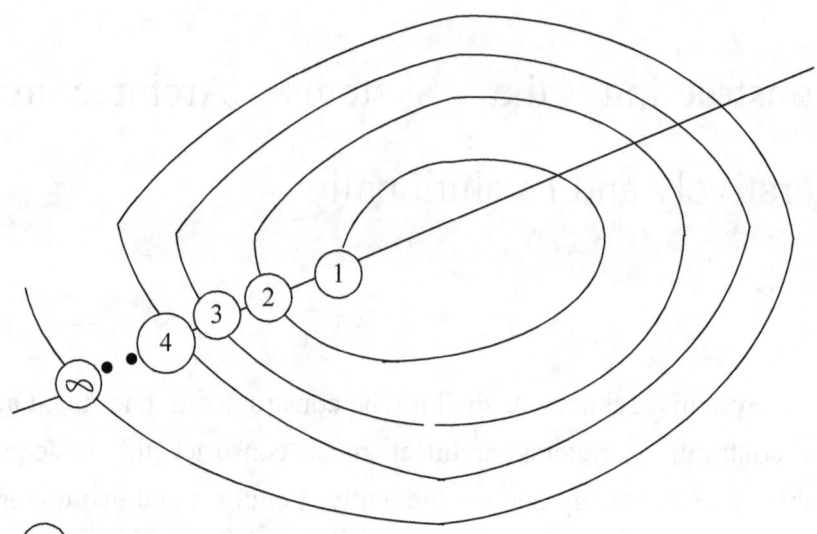

(1) : Systems Architecture Version 1

(2) : Systems Architecture Version 2

(3) : Systems Architecture Version 3

(4) : Systems Architecture Version 4

• • •

(∞) : Systems Architecture Version ∞

Systems architecture *version n* is sometimes referred to as the baseline (As-Is) architecture which represents the current system that has been formally reviewed and agreed upon. On the other hand, systems architecture *version n+1* is sometimes referred to as the target (To-Be) architecture which represents the goal system that will be formally constructed.

Architecture Development Method

If we adopt the iterative and evolutional construction of systems architecture approach, then we would obtain the architecture development method (ADM) as shown below.

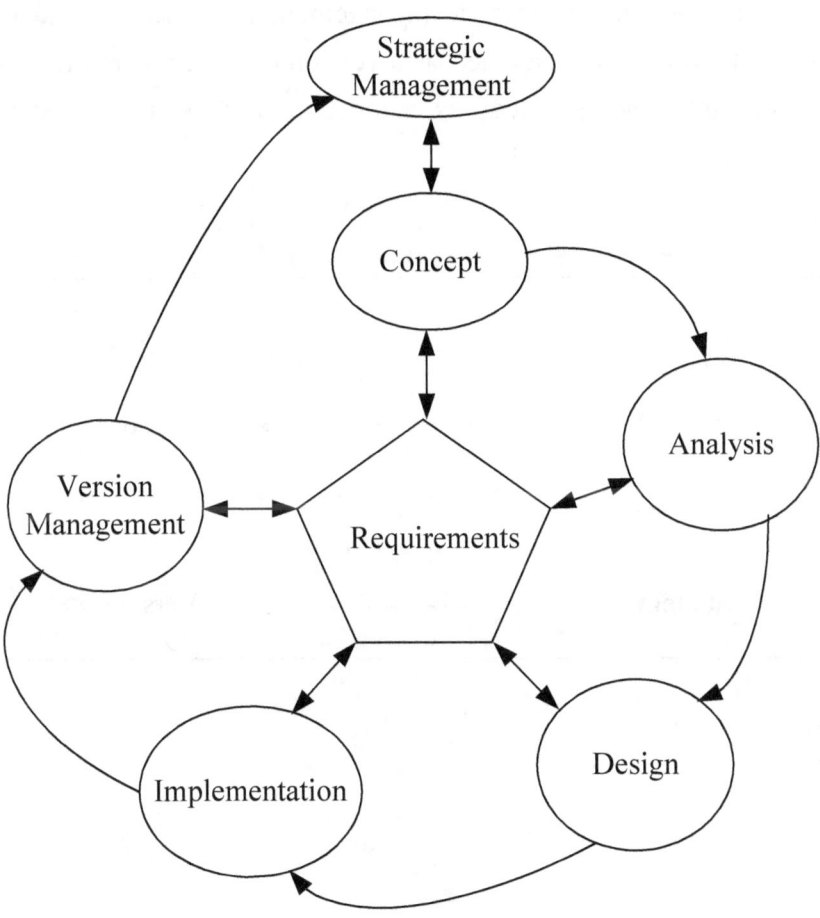

The iterative and cyclic ADM, being utilized by a systems architect to accomplish each version management of systems architecture, shall do the strategic management first and then go through the concept, analysis, design and implementation phases of systems architecture construction. Every phase checks with the requirements to make sure that each version of the constructed systems architecture is what the users want.

The output of strategic management is a strategy and the output of version management is a version of systems architecture. Accordingly, each strategy is mapped to a version of systems architecture as shown below.

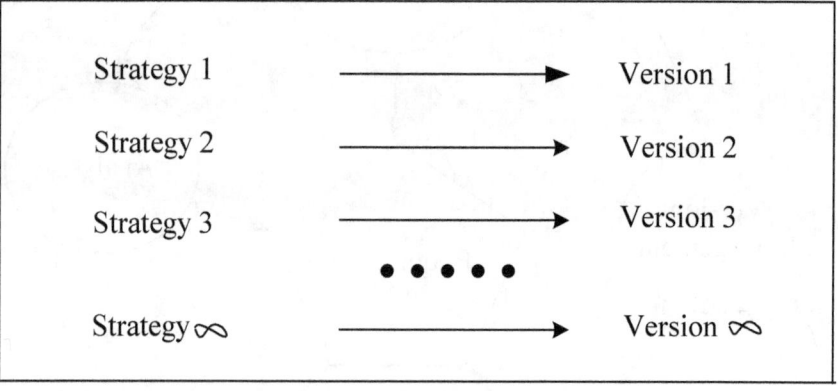

View Model

A system comprises multiple views such as strategy/version n, strategy/version n+1, concept, analysis, design, implementation, structure, behavior and input/output data views. We can represent all these multiple views in a one-dimensional array as shown below.

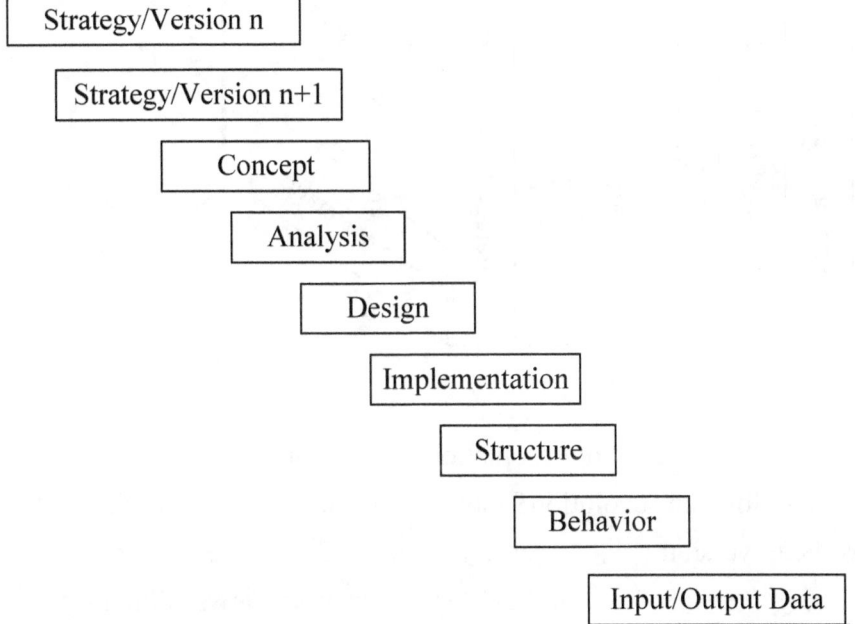

We can also describe and represent all these multiple views in the above figure in a three-dimensional matrix as shown below.

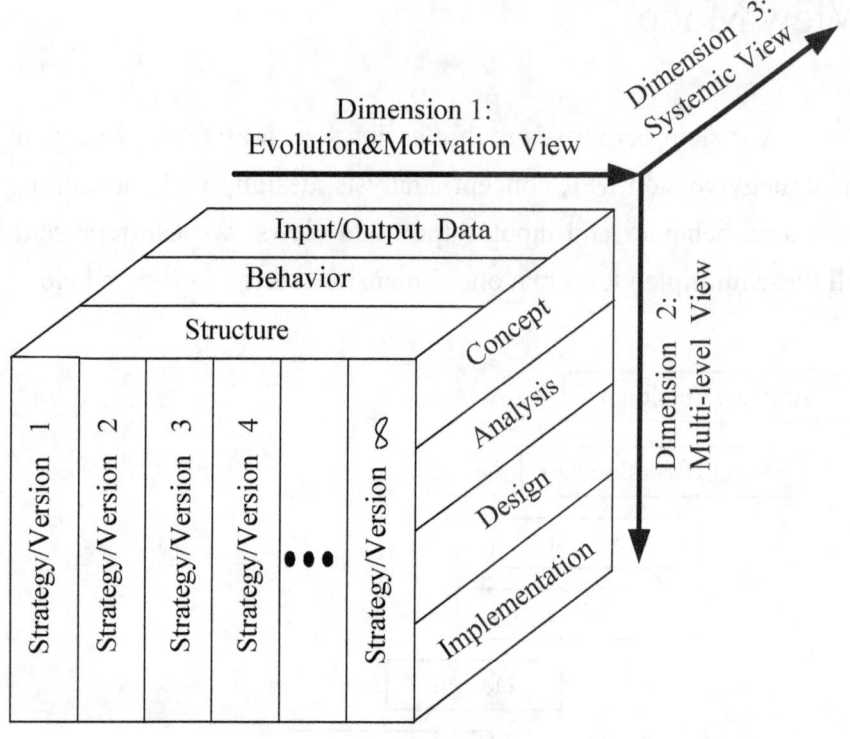

In the matrix representation of multiple views, dimension 1 stands for the evolution&motivation view which contains the strategy/version 1, strategy/version 2, strategy/version 3, strategy/version 4,…, and strategy/version ∞ views; dimension 2 stands for the multi-level (hierarchical) view which contains the concept, analysis, design and implementation views; dimension 3 stands for the systemic view which contains the structure, behavior, input/output data views. The matrix representation of multiple views is also called a view model (VM) or architecture framework (AF).

PART IV: STRUCTURE-BEHAVIOR COALESCENCE FOR SYSTEMS ARCHITECTURE

Multiple Views Coalescence to Achieve the Systems Architecture

Systems architecture has been defined as a coalescence model of multiple views. Multiple views coalescence uses only a single coalescence model. Strategy/version n, strategy/version n+1, concept, analysis, design, implementation, structure, behavior and input/output data views are all integrated in this MVC architecture.

Generally, MVC architecture is synonymous with the systems architecture. In other words, multiple views coalescence sets a path to achieve the systems architecture.

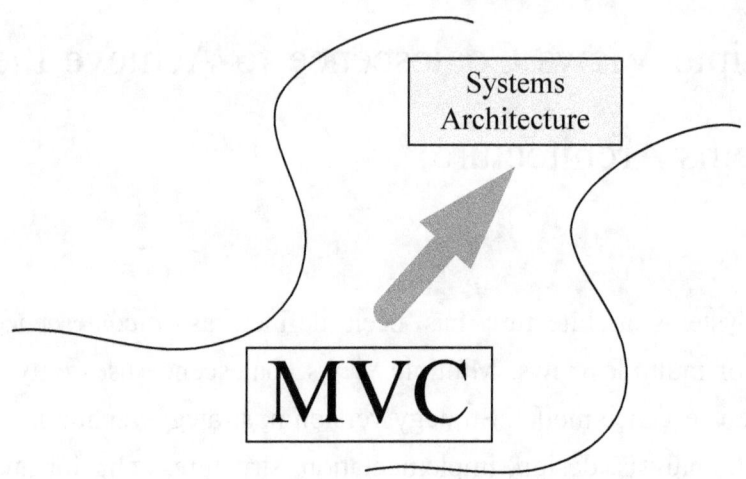

In the MVC architecture, multiple views must be attached to or built on the systems structure. In other words, multiple views shall not exist alone; they must be loaded on the systems structure just like a cargo is loaded on a ship. There will be no multiple views if there is no systems structure. Stand-alone multiple views are not meaningful.

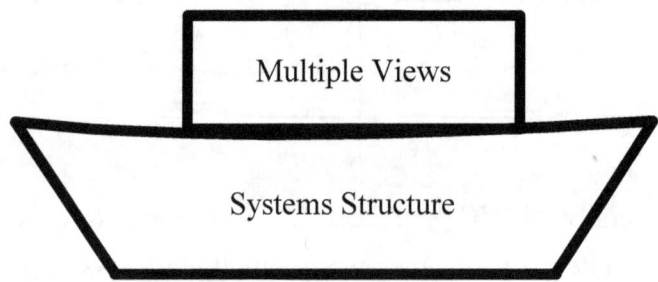

Integrating the Systems Structures and Systems Behaviors

By integrating the systems structure and systems behavior, we obtain structure-behavior coalescence (SBC) within the system.

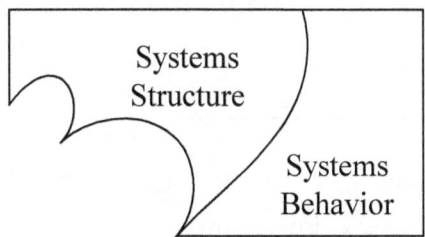

Structure-behavior coalescence has never been used in any systems model (SM) for systems development. There are many advantages to use the structure-behavior coalescence approach to integrate the systems structure and systems behavior.

SBC architecture uses a single coalescence model. Systems structures and systems behaviors are integrated in this SBC architecture.

Since systems structures and systems behaviors are so tightly integrated, we sometimes claim that the core theme of SBC architecture is: "Systems Architecture = Systems Structure + Systems Behavior."

Systems Architecture	=	**Systems Structure**	+	**Systems Behavior**

So far, systems behaviors are separated from systems structures in most cases. For example, the well-known structured systems analysis and design (SSA&D) approach uses structure charts (SC) to represent the systems structure and data flow diagrams (DFD) to represent the systems behavior. SC and DFD are two heterogeneous and separated models. They are so separated like that there is "Pacific Ocean" between them.

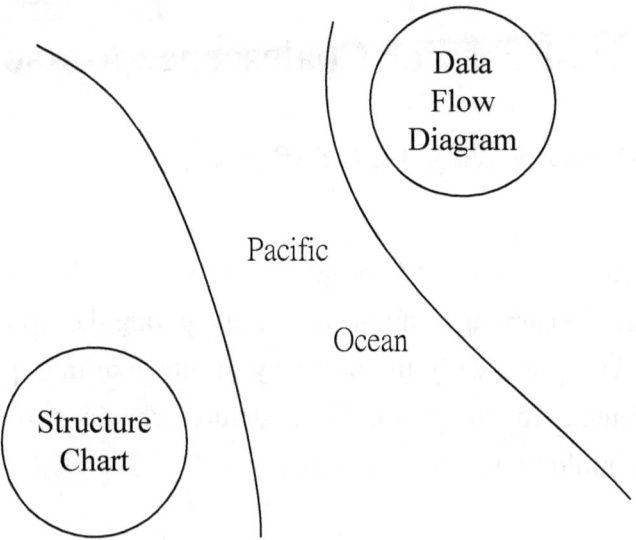

Pacific

Ocean

Structure-Behavior Coalescence to Facilitate Multiple Views Coalescence

Since structure and behavior views are the two most prominent ones among multiple views, integrating the structure and behavior views is clearly the best way to integrate multiple views of a system. In other words, structure-behavior coalescence facilitates multiple views coalescence.

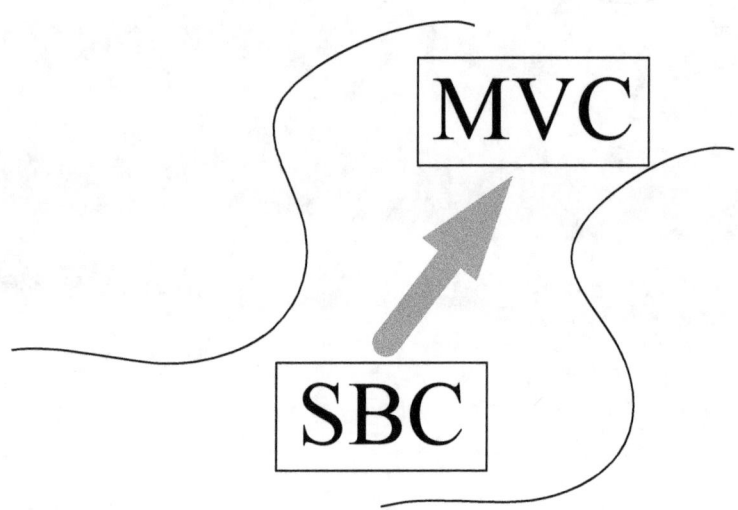

Structure-Behavior Coalescence to Achieve the Systems Architecture

We declared that multiple views coalescence sets a path to achieve the desired systems architecture with the most efficient approach. We also declared that structure-behavior coalescence facilitates multiple views coalescence.

Combining the above two declarations, we conclude that structure-behavior coalescence sets a path to achieve the systems architecture. In this case, SBC architecture is also synonymous with the systems architecture.

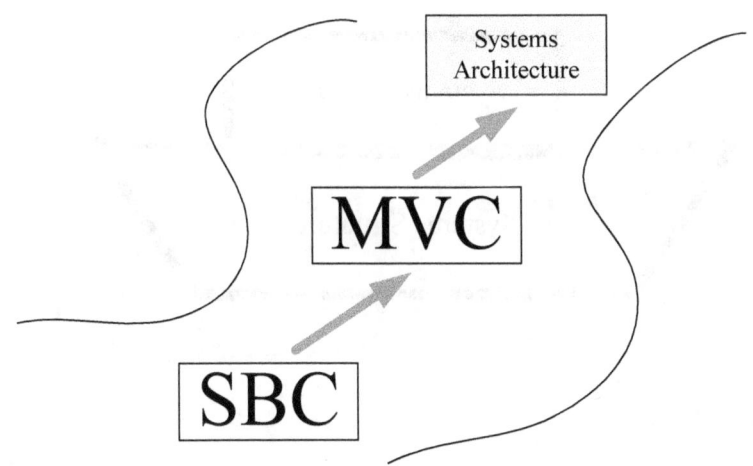

SBC architecture strongly demands that the structure and behavior views must be coalesced and integrated. This never happens in other architectural approaches such as Zachman Framework, The Open Group Architecture Framework (TOGAF), Department of Defense Architecture Framework (DoDAF) and Unified Modeling Language (UML). Zachman Framework does not offer any mechanism to integrate the structure and behavior views. TOGAF, DoDAF and UML do not, either.

In the SBC architecture, a systems behavior must be attached to or built on a systems structure. In other words, a systems behavior can not exist alone; it must be loaded on a systems structure just like a cargo is loaded on a ship. There will be no systems behavior if there is no systems structure. A stand-alone systems behavior is not meaningful.

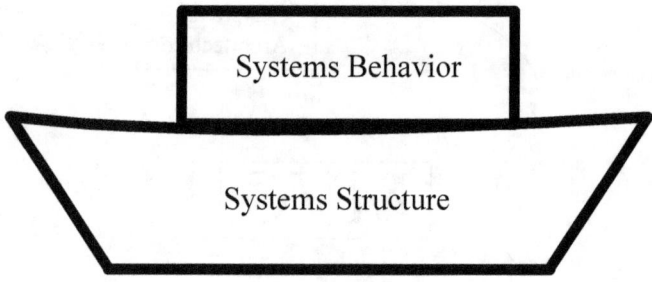

PART V: INTRODUCTION TO SBC ARCHITECTURE

Definition of SBC Architecture

Here, let us first give the SBC architecture a definition as shown below.

SBC architecture,
through structure-behavior coalescence,
truly is an integrated whole of a system's multiple views, i.e., structure, behavior, and other views, embodied in its components, their interactions with each other and the environment, and the principles and guidelines governing its design and evolution.

From the above definition, we find out that SBC architecture, through structure-behavior coalescence, is a truly integrated whole of a system's multiple views, i.e., structure, behavior and other views, embodied in its assembled components, their interactions (or handshakes) with each other and the environment, and the principles and guidelines governing its design and evolution.

SBC architecture includes: a) SBC architecture description language (SBC-ADL), b) SBC architecture development method (SBC-ADM) and c) SBC view model (SBC-VM).

SBC Architecture Description Language

SBC-ADL uses six fundamental diagrams to describe the integration of systems structure and systems behavior of a system. These diagrams are: a) architecture hierarchy diagram (AHD), b) framework diagram (FD), c) component operation diagram (COD), d) component connection diagram (CCD), e) structure-behavior coalescence diagram (SBCD) and f) interaction flow diagram (IFD).

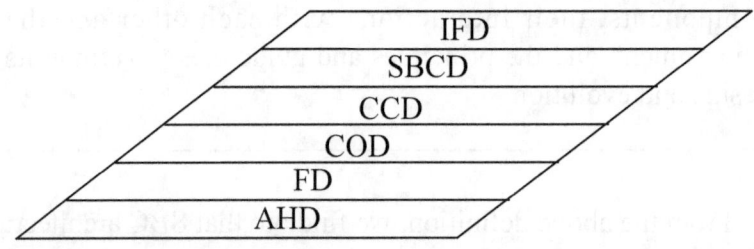

SBC-ADL uses AHD, FD, COD, CCD, SBCD and IFD to depict the systems structure and systems behavior of a system.

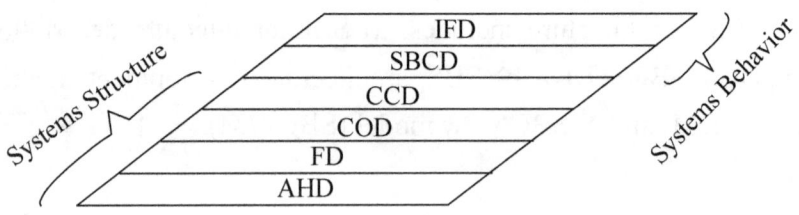

SBC Architecture Development Method

The term of architecture development method (ADM) is first used in the open group architecture framework. Through the iterative and cyclic ADM, a systems architect is able to construct the systems architecture smoothly.

SBC architecture development method (SBC-ADM), being utilized by a systems architect to accomplish each version management of the systems architecture, shall do the strategic management first and then go through the concept, analysis, design and implementation phases of systems architecture construction. Every phase shall check with the requirements to make sure that each version of the constructed systems architecture is what the users want.

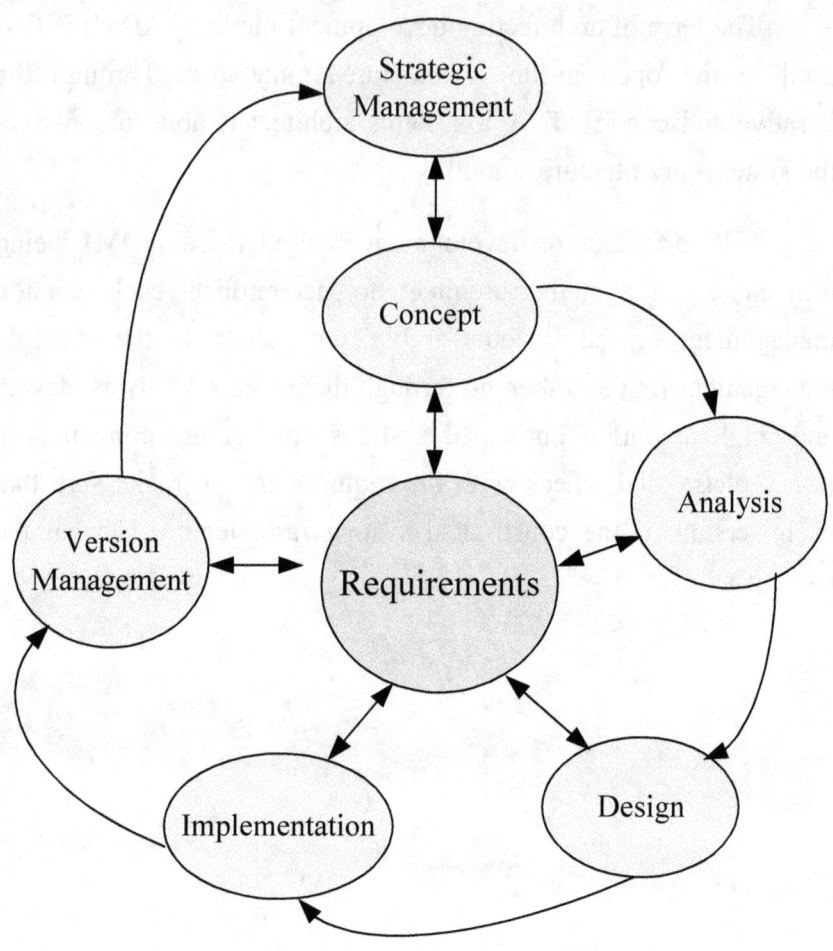

The output of strategic management is a strategy and the output of version management is a version of systems architecture. Accordingly, each strategy is mapped to a version of systems architecture.

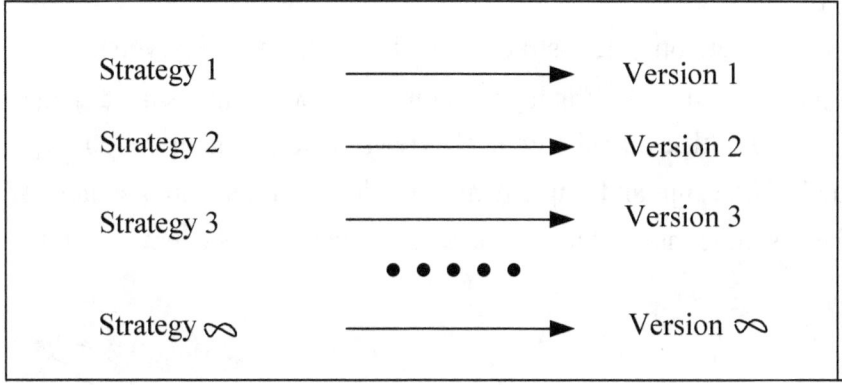

SBC View Model

SBC view model (SBC-VM) is a three-dimensional matrix representation of a system's multiple views. Dimension 1 stands for the evolution&motivation view which contains the strategy/version 1, strategy/version 2, strategy/version 3, strategy/version 4, strategy/version ∞ views; dimension 2 stands for the multi-level (hierarchical) views which contain the concept, analysis, design and implementation views; dimension 3 stands for the systemic view which contains the structure and behavior views.

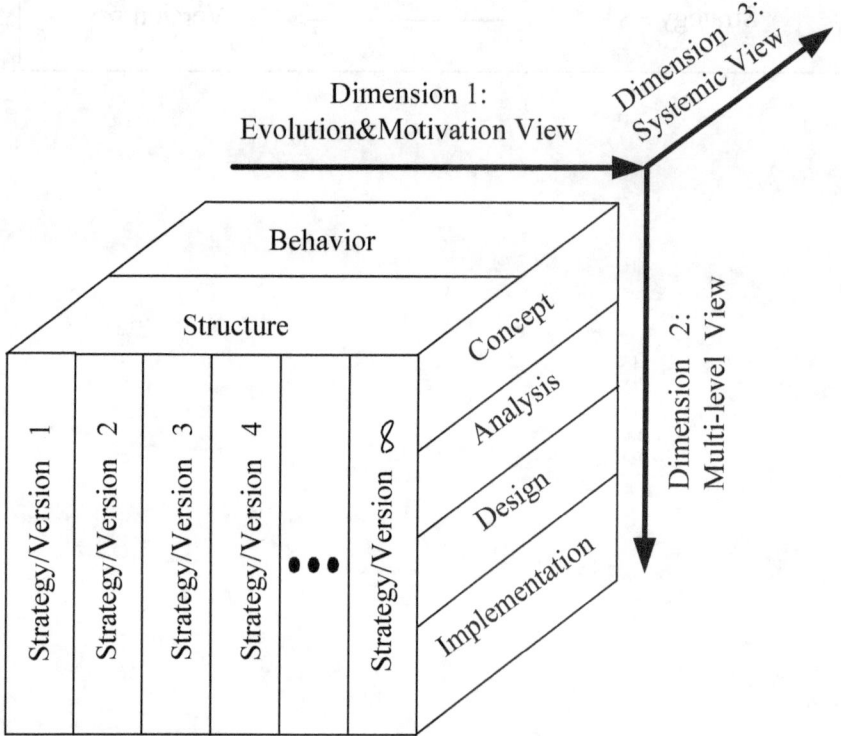

According to the approach of SBC architecture description language (SBC-ADL), the structure view consists of AHD, FD, COD and CCD; the behavior view consists of SBCD and IFD. Also, FD consists of business layer, application layer, data layer and technology layer. Adding these ideas, we then get the complete SBC view model.

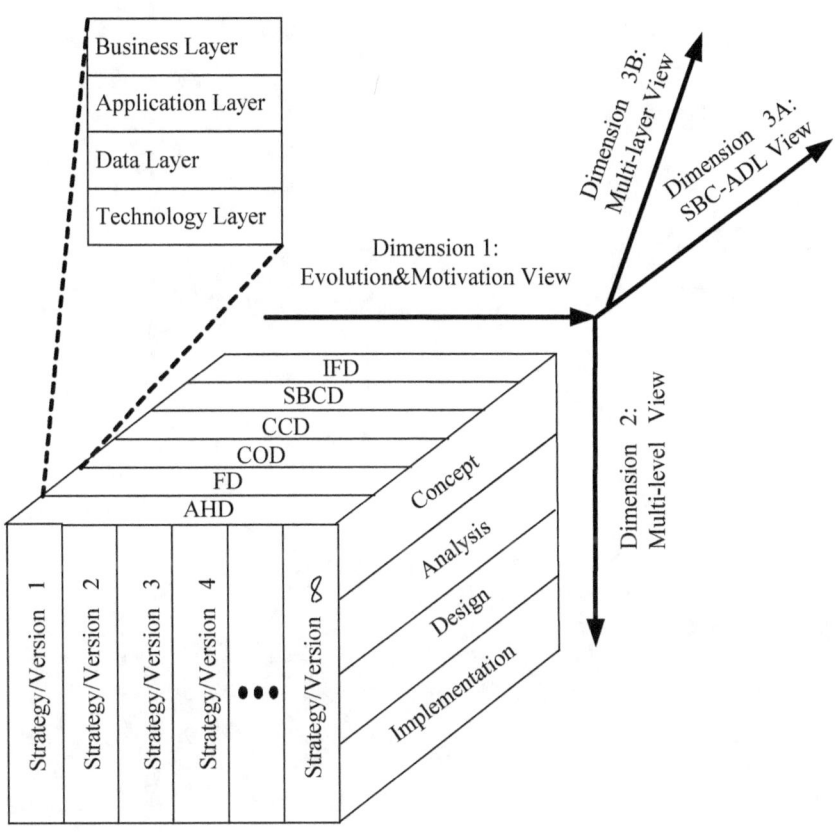

PART VI: INTRODUCTION TO GENERAL SYSTEMS THEORY 2.0

General Systems Theory 2.0 Defining a System

SBC architecture provides an elegant way to integrate the structure and behavior of a system. Therefore, general systems theory 2.0 shall use the SBC architecture to define a system. A system is redefined, by general systems theory 2.0, as follows.

> A system,
> through the SBC architecture,
> truly is an integrated whole,
> embodied in its components,
> their interactions with each other and the environment,
> and the principles and guidelines governing its design and evolution.

Since general systems theory 2.0 uses the SBC architecture to define a system, general systems theory 2.0 is also called general architectural theory.

General Systems Theory 2.0

General Architectural Theory

According to the above description, general systems theory 2.0 uses the SBC architecture to define a system. Based on the SBC view model, there are three significant dimensions for general systems theory 2.0 to define a system.

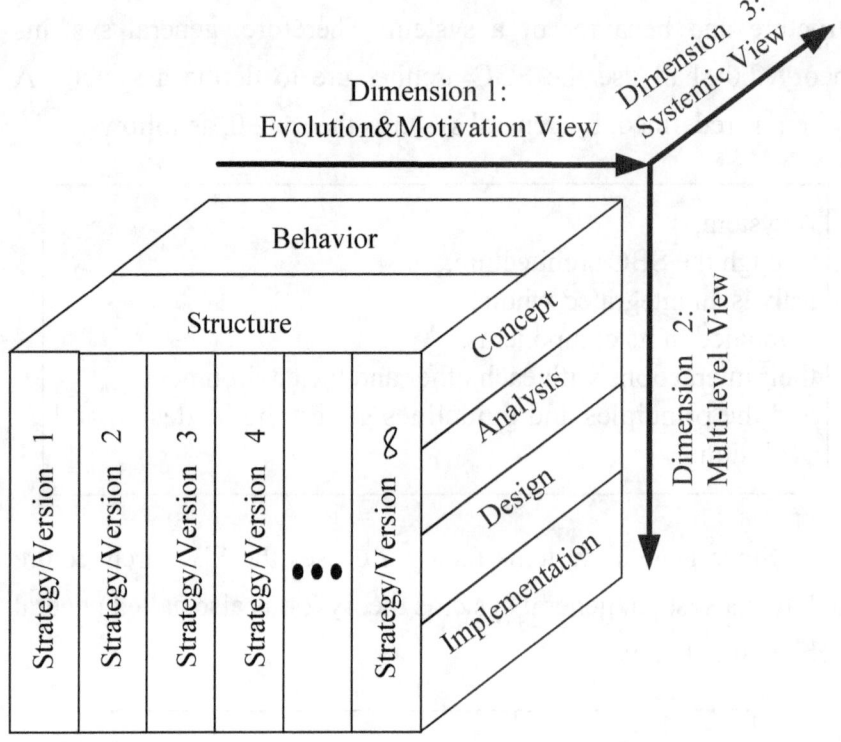

Dimension 1 stands for the evolution&motivation view which contains the strategy/version 1, strategy/version 2, strategy/version 3, strategy/version 4,···, and strategy/version ∞ views. Dimension 2 stands for the multi-level (hierarchical) view which contains the concept, analysis, design and implementation views. Dimension 3 stands for the systemic view which contains the structure and behavior views.

Evolution&Motivation View

A system, not matter it is physical or virtual, will always change from time to time. A system evolves when it changes. Evolution of a system is shown below.

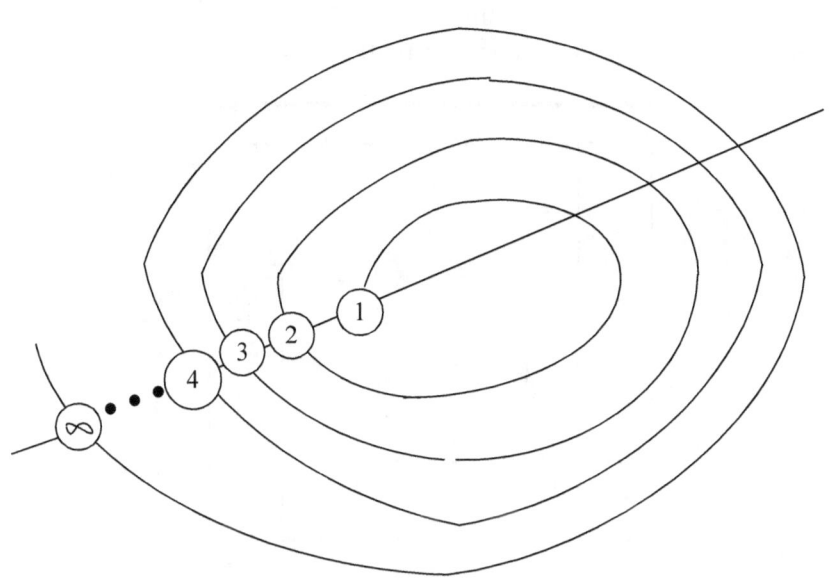

(1) : Systems Definition Version 1

(2) : Systems Definition Version 2

(3) : Systems Definition Version 3

(4) : Systems Definition Version 4

• • •

(∞) : Systems Definition Version ∞

Each time when a system changes or evolves, we shall get a new version of its systems definition. In the above figure, *version 1* stands for the original systems definition of a system and evolves into *version 2*, *version 3*, *version 4,…*, and *version* ∞ gradually.

Evolution of a system is represented by the SBC architecture, as the evolution&motivation view.

Dimension 1:
Evolution&Motivation View

Strategy/Version 1 | Strategy/Version 2 | Strategy/Version 3 | Strategy/Version 4 | ••• | Strategy/Version ∞

In the evolution&motivation view, we see that each strategy is mapped to a version of systems definition.

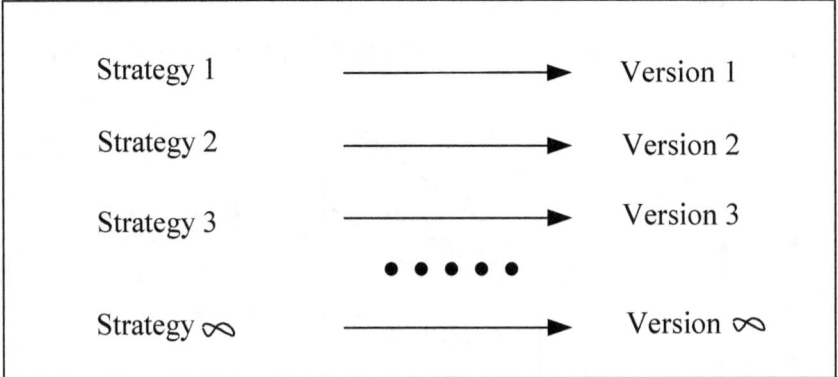

86

Multi-Level View

In the SBC architecture, multi-level (hierarchical) view contains the concept, analysis, design and implementation views.

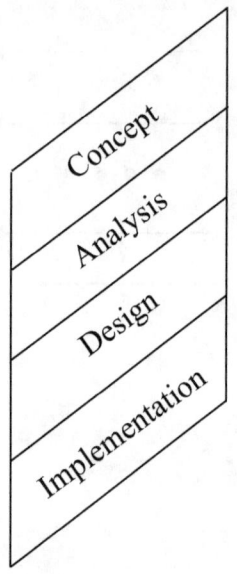

Concept view corresponds to an executive summary for an administrator who wants an estimate of the scope of the system, what it would cost, and how it would relate to the general environment in which it will operate. Analysis view corresponds to a summary for an analyzer who works on the analysis of a system. Analysis view is one level down structural decomposition (with observation congruence verification) of the concept view. Design view describes what a designer has accomplished for his task. Design view is one level down structural decomposition (with observation congruence verification) of the analysis view.

Implementation view shows what an implementer has done for his work. Implementation view is one level down structural decomposition (with observation congruence verification) of the design view.

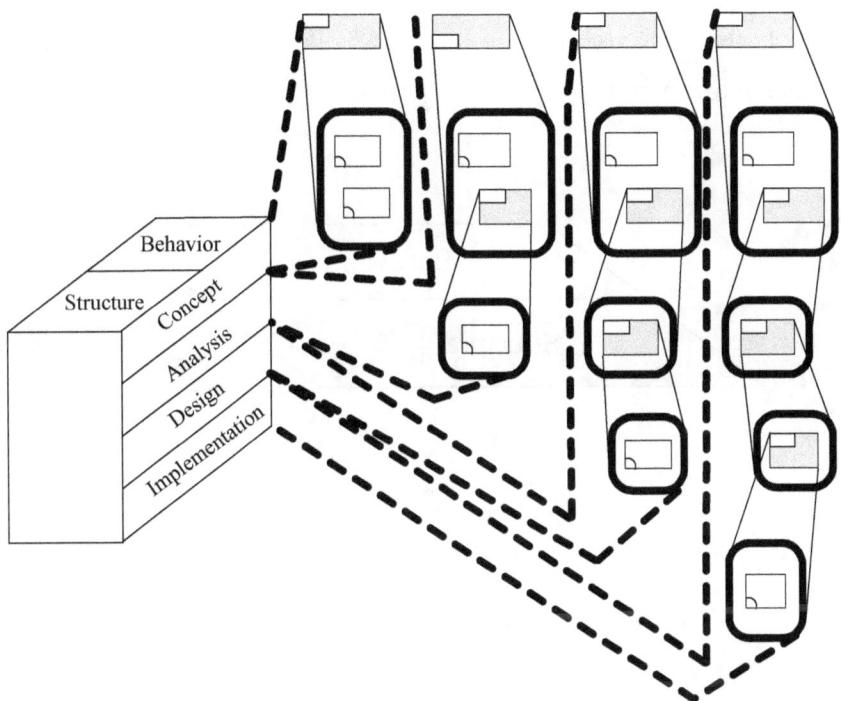

Systemic View

In the SBC architecture, systemic view contains the structure and behavior views.

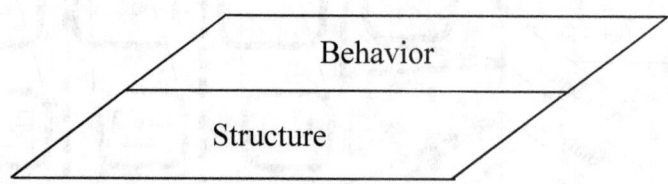

According to the approach of SBC architecture description language (SBC-ADL), the structure view consists of AHD, FD, COD and CCD; the behavior view consists of SBCD and IFD. Also, FD consists of business layer, application layer, data layer and technology layer. Adding these ideas to the above figure, we then get the systemic view of general systems theory 2.0 (general architectural theory), as shown below.

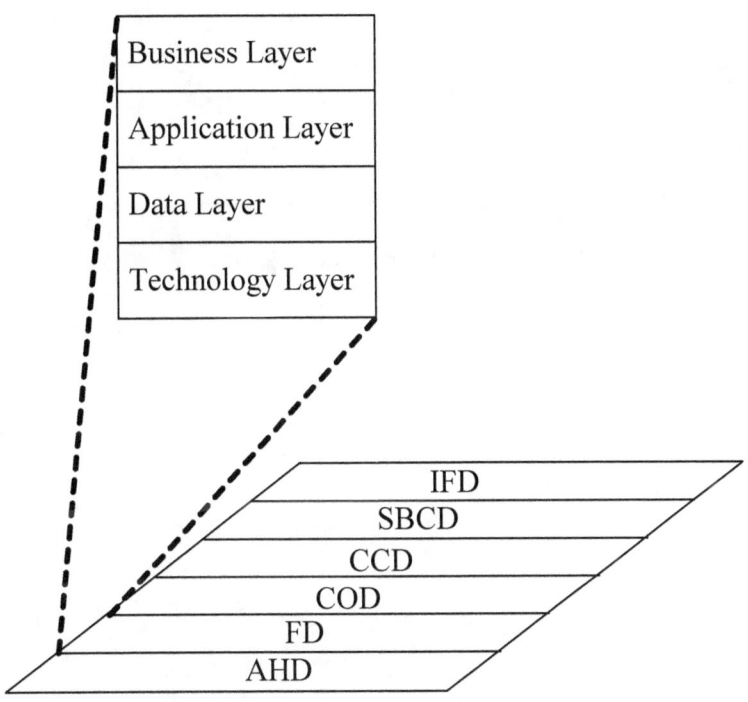

PART VII: SYSTEMIC VIEW OF GENERAL SYTEMS THEORY 2.0

Architecture Hierarchy Diagram

General systems theory 2.0 uses an architecture hierarchy diagram (AHD) to define the multi-level decomposition and composition of a system.

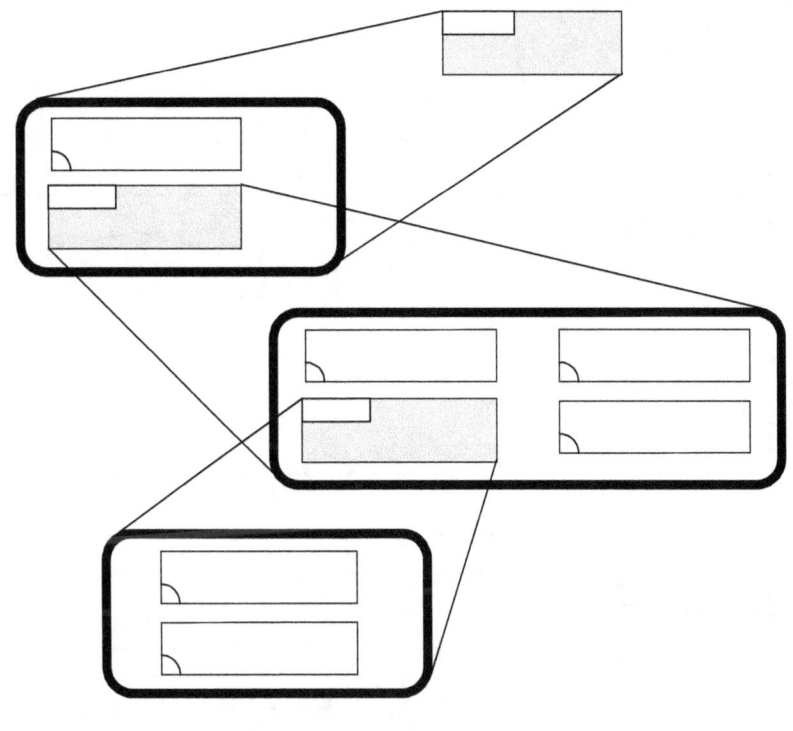

: Aggregated System

: Non-Aggregated System, Component

Framework Diagram

General systems theory 2.0 uses a framework diagram (FD) to define the multi-laye (also referred to as multi-tier) decomposition and composition of a system.

: Component

Component Operation Diagram

General systems theory 2.0 uses the component operation diagram (COD) to define all components' operations of a system.

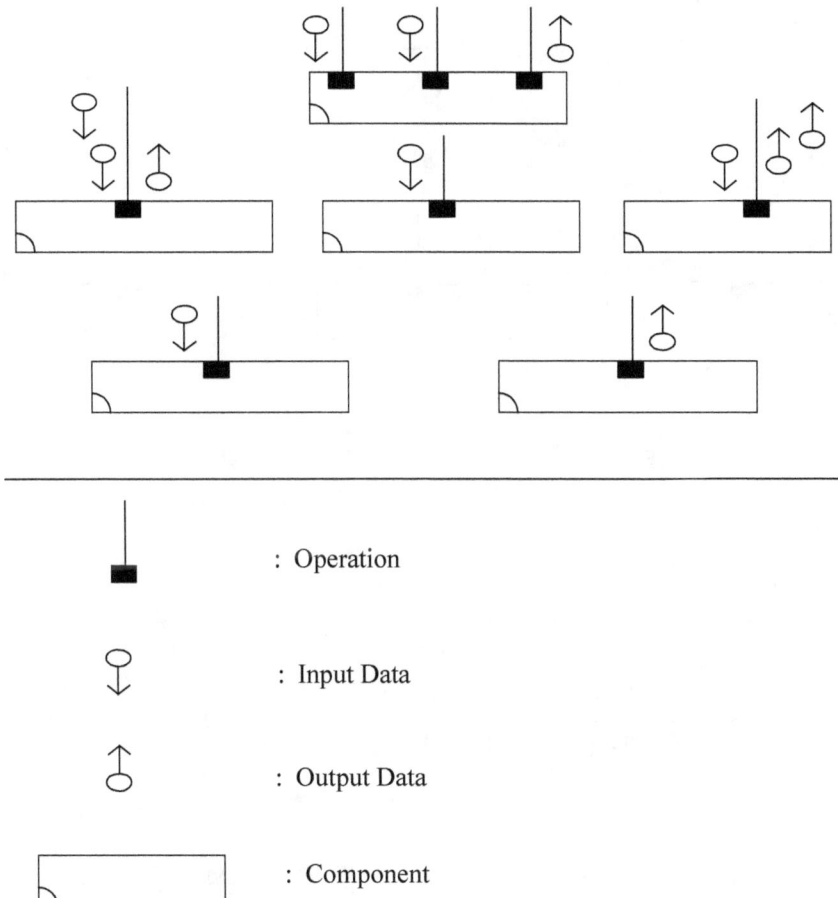

: Operation

: Input Data

: Output Data

: Component

Component Connection Diagram

General systems theory 2.0 uses a component connection diagram (CCD) to define how all components and actors are connected within a system.

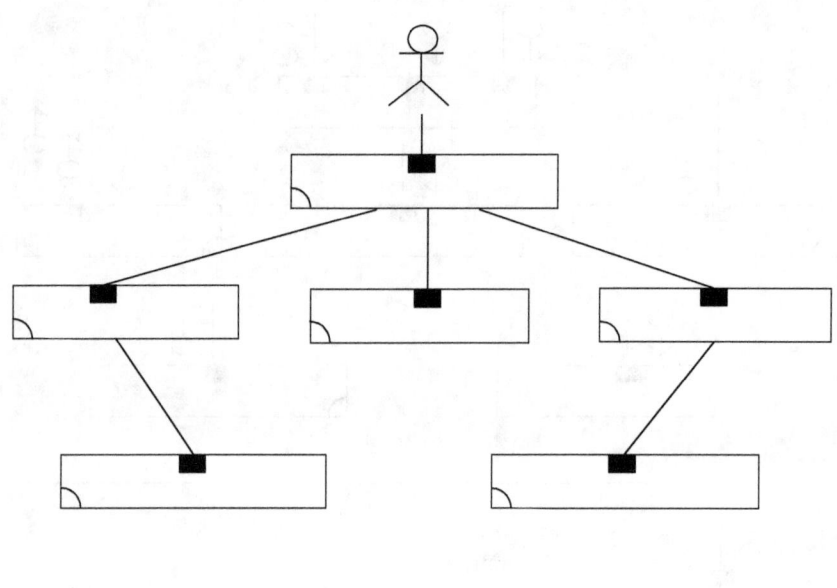

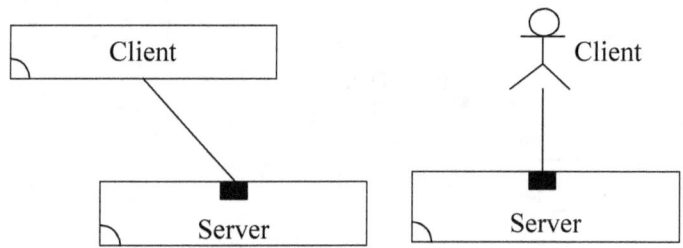

Structure-Behavior Coalescence Diagram

General systems theory 2.0 uses a structure-behavior coalescence diagram (SBCD) to define the systems structure and systems behavior coexisting in a system.

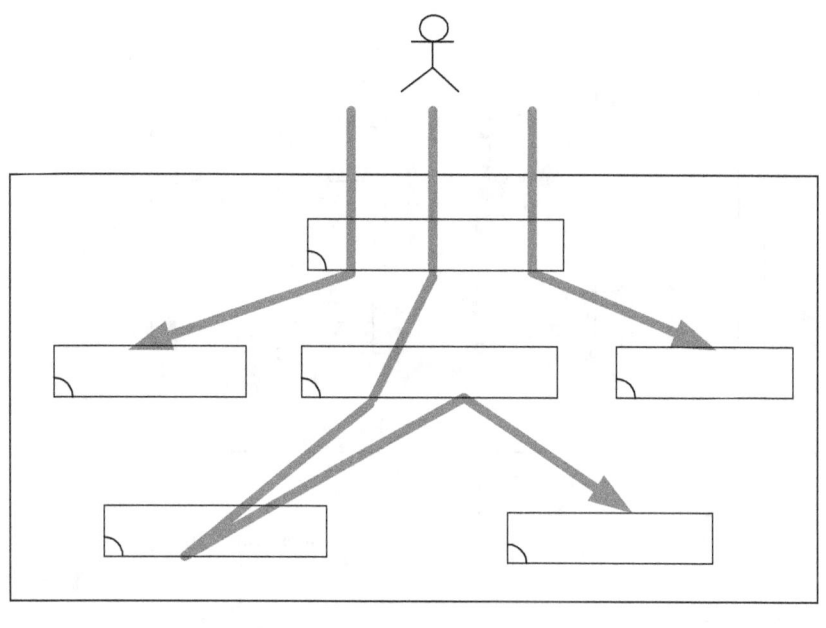

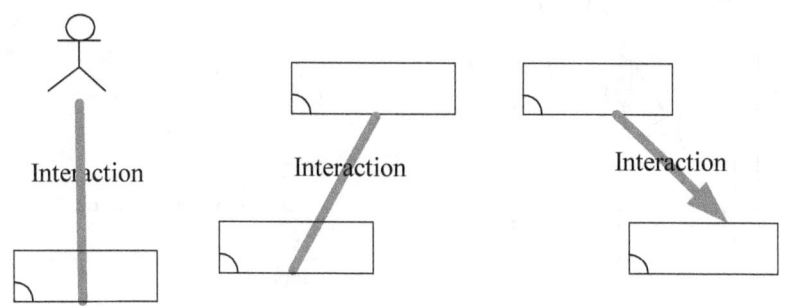

Interaction Flow Diagram

General systems theory 2.0 uses an interaction flow diagram (IFD) to define each individual behavior of the overall behavior of a system.

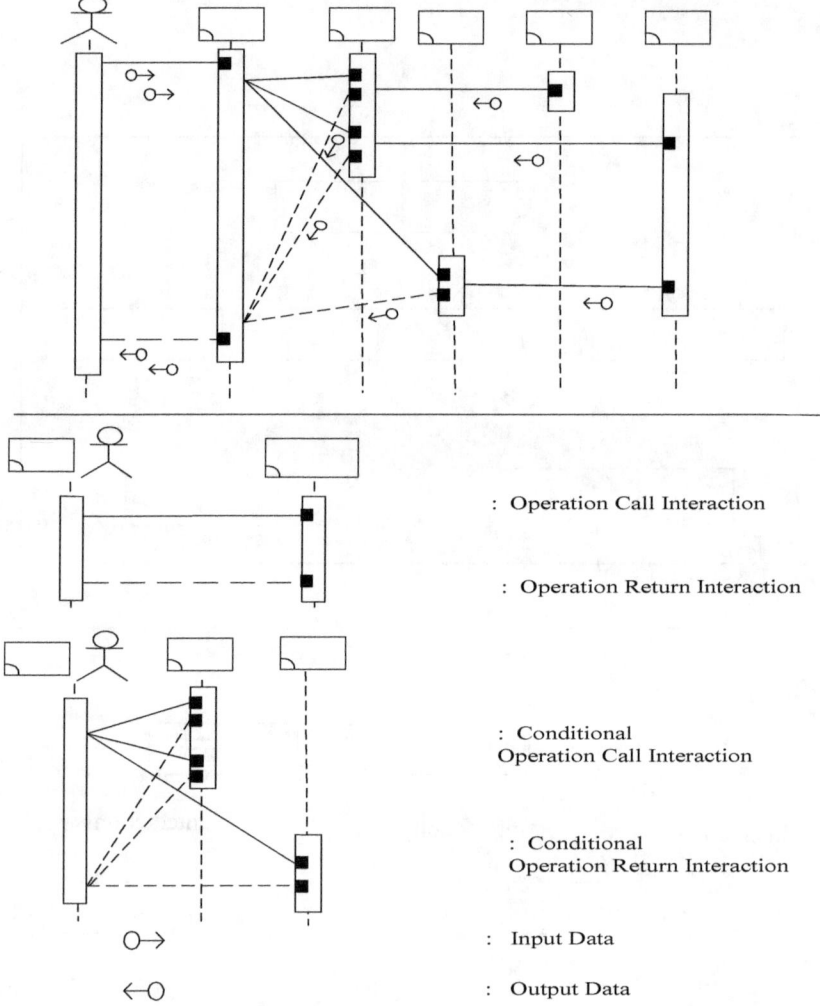

: Operation Call Interaction

: Operation Return Interaction

: Conditional
Operation Call Interaction

: Conditional
Operation Return Interaction

O→ : Input Data

←O : Output Data

PART VIII: MULTI-LEVEL VIEW OF GENERAL SYSTEMS THEORY 2.0

Concept View

Concept view corresponds to an executive summary for an administrator who wants an overview or estimate of the scope of the system, what it would cost, and how it would relate to the general environment in which it will operate.

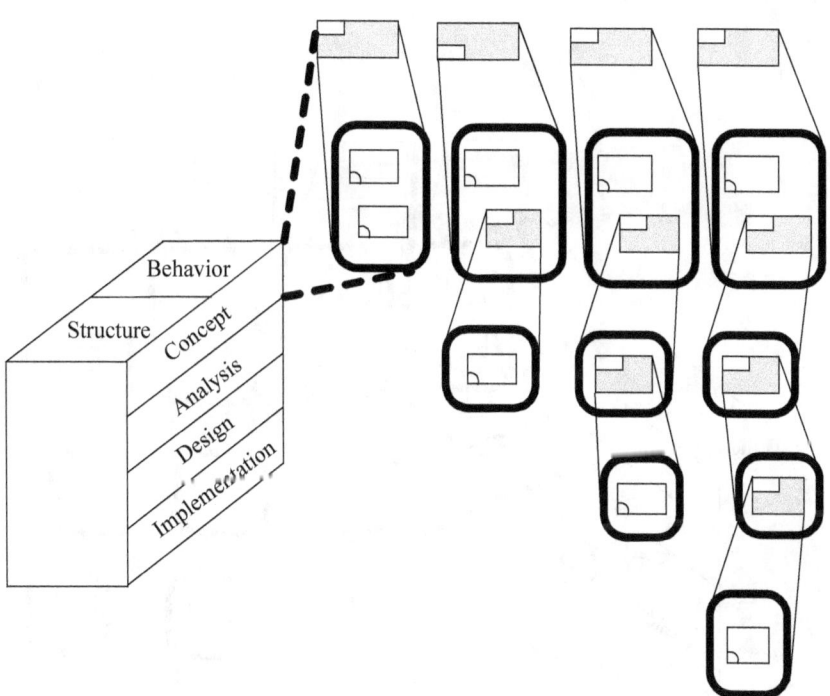

Analysis View

Analysis view corresponds to a summary for an analyzer who works on the analysis of a system. The analysis is mainly to find out what the system is. When working on the analysis, we only ask what this system is about, but may not provide sufficient focus on how the system is actually designed.

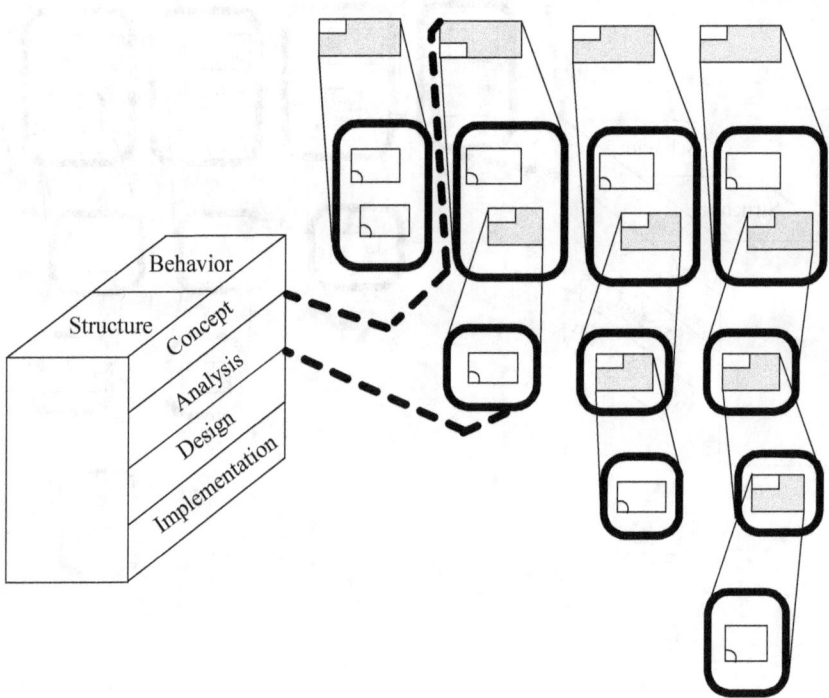

Analysis view is one level down structural decomposition (with observation congruence verification) of the concept view. That is, we shall not create the analysis view from the scratch. Instead, we will construct the analysis view by decomposing the concept view.

Design View

Design view describes what a designer has accomplished for his work.

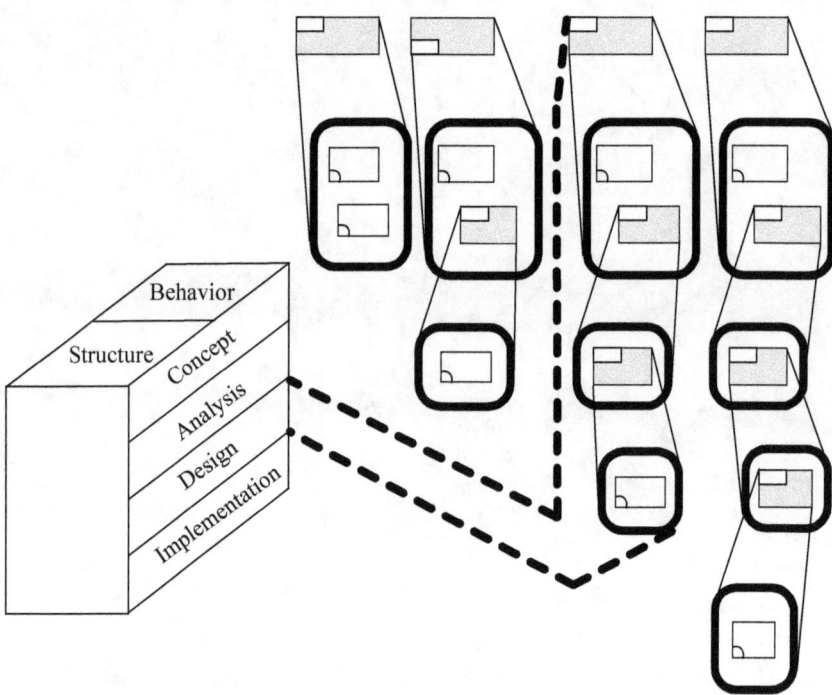

Design view is one level down structural decomposition (with observation congruence verification) of the analysis view. That is, we shall not create the design view from the scratch. Instead, we will construct the design view by decomposing the analysis view.

Implementation View

Implementation view shows what an implementer has done for his work.

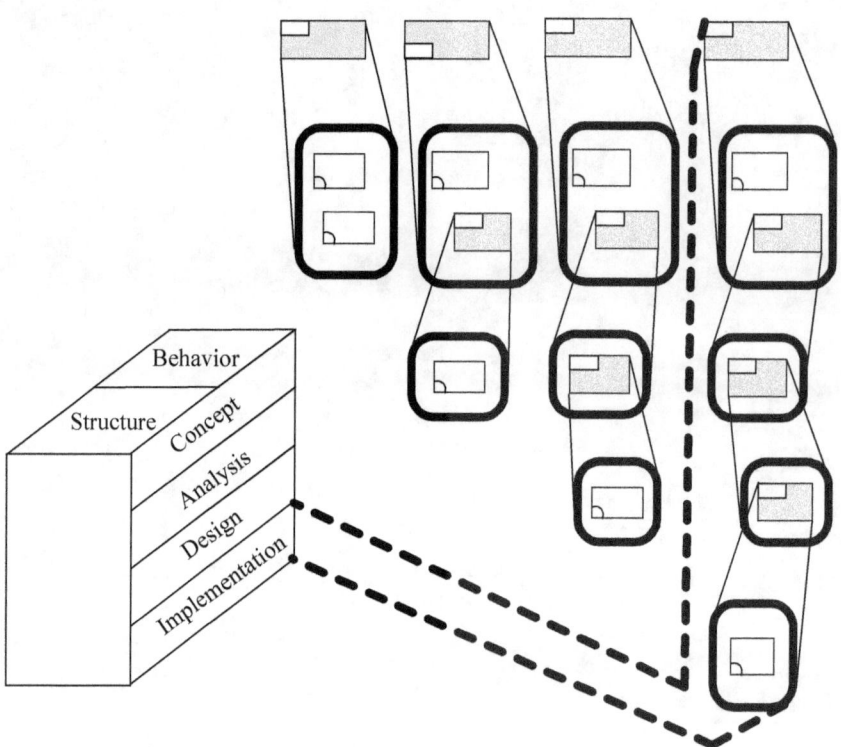

Implementation view is one level down structural decomposition (with observation congruence verification) of the design view. That is, we shall not create the implementation view from the scratch. Instead, we will construct the implementation view by decomposing the design view.

PART IX:
EVOLTION/MOTIVATION VIEW
OF GENERAL SYSTEMS THEORY
2.0

Higher-Order Systems

Higher-order systems interact with the environment through the exchange of not only matter, energy, data, information, or message but also systems.

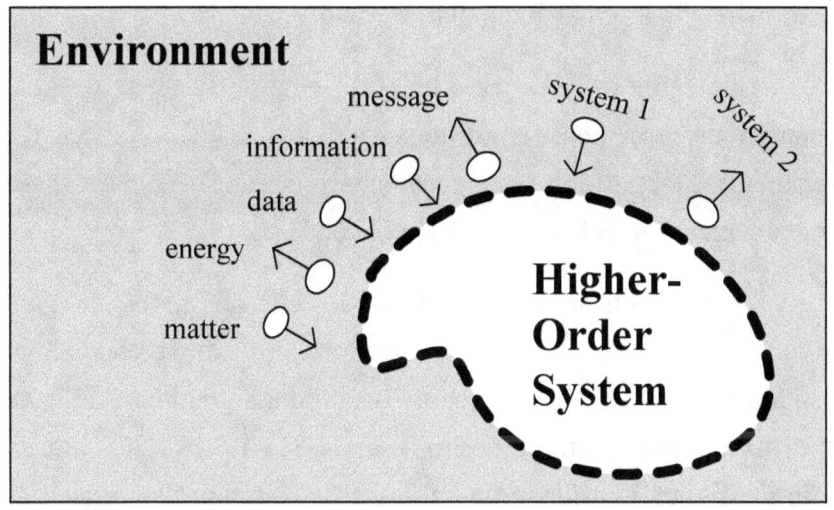

For a higher-order system is computable, it must be monotonic and continuous.

Strategic Means

In the business motivation model, strategy is the human attempt to get to "desirable ends with available means". Strategic means analyzes the major initiatives taken by a company's top management on behalf of business owners, involving resources and performance in internal and external environments.

Strategic means include: (a) goal drivers, (b) goal assumptions, (c) goal constraints and (d) SWOT (strengths, weaknesses, opportunities, threat) analysis, etc. We use these strategic means to achieve the desirable ends.

Goal drivers are up from the policy considerations, the goal driver is kind of why we want to have those desirable ends. Goal assumptions are taking into account of those assumptions that have a positive impact on these desirable ends. Goal constraints are up from the policy considerations, the goal constraints are related to those restrictions which have a negative impact on those desirable ends. SWOT analysis is to analyze the internal strengths, weaknesses, opportunities and threats, and so for executing this strategy.

Motivation Model is a Higher-Order System

Motivation model is a higher-order system. Motivation model will output a system (goal) for each strategy.

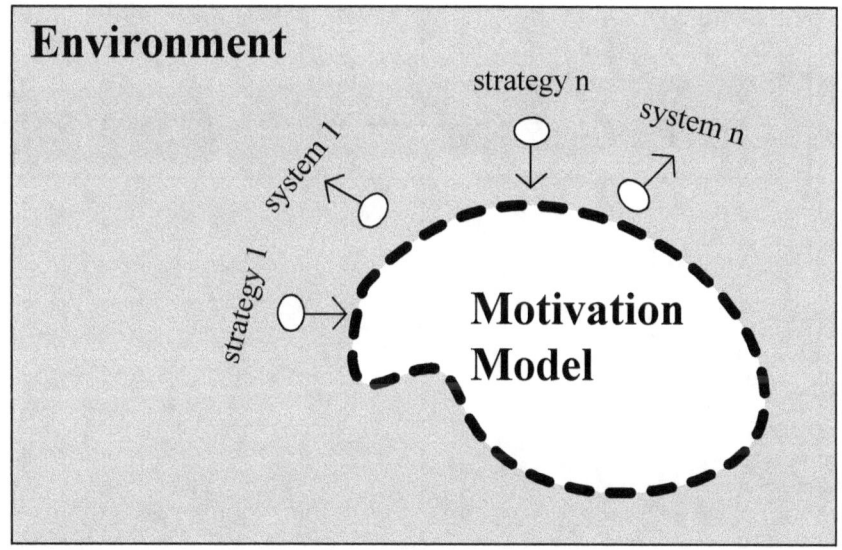

One strategy is mapped to one version for any systems definition.

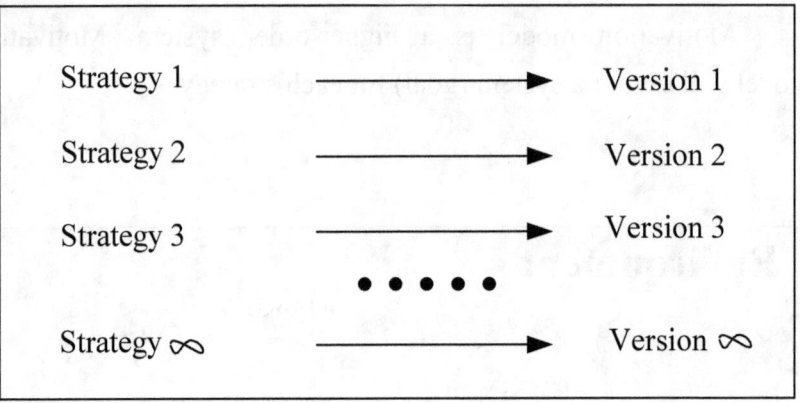

Evolution&Motivation View

Evolution of a system is represented, by general systems theory 2.0, as the SBC evolution&motivation view. For any systems definition, one strategy is mapped to one version.

Evolution&Motivation View

www.ingramcontent.com/pod-product-compliance
Lightning Source LLC
Chambersburg PA
CBHW060350190526
45169CB00002B/547